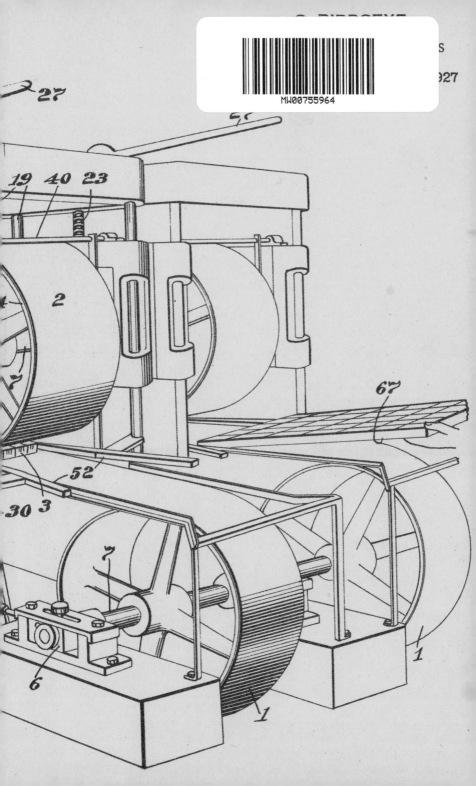

BIRDSEYE

Also by Mark Kurlansky

NONFICTION

*What? Are These the 20 Most Important Questions in Human History—
or Is This a Game of 20 Questions?*

Hank Greenberg: The Hero Who Didn't Want to Be One

*The Eastern Stars: How Baseball Changed the Dominican Town of
San Pedro de Macorís*

*The Last Fish Tale: The Fate of the Atlantic and Survival in Gloucester,
America's Oldest Fishing Port and Most Original Town*

The Big Oyster: History on the Half Shell

Nonviolence: Twenty-Five Lessons from the History of a Dangerous Idea

1968: The Year That Rocked the World

Salt: A World History

The Basque History of the World

Cod: A Biography of the Fish That Changed the World

A Chosen Few: The Resurrection of European Jewry

A Continent of Islands: Searching for the Caribbean Destiny

FICTION

Edible Stories: A Novel in Sixteen Parts

Boogaloo on 2nd Avenue: A Novel of Pastry, Guilt, and Music

The White Man in the Tree, and Other Stories

TRANSLATION

The Belly of Paris, by Émile Zola

ANTHOLOGIES

*The Food of a Younger Land: A Portrait of American Food—
Before the National Highway System, Before Chain Restaurants,
and Before Frozen Food, When the Nation's Food Was Seasonal, Regional,
and Traditional—from the Lost WPA Files*

*Choice Cuts: A Savory Selection of Food Writing from
Around the World and Throughout History*

CHILDREN/YOUNG ADULT

Battle Fatigue

World Without Fish

The Story of Salt

The Girl Who Swam to Euskadi

The Cod's Tale

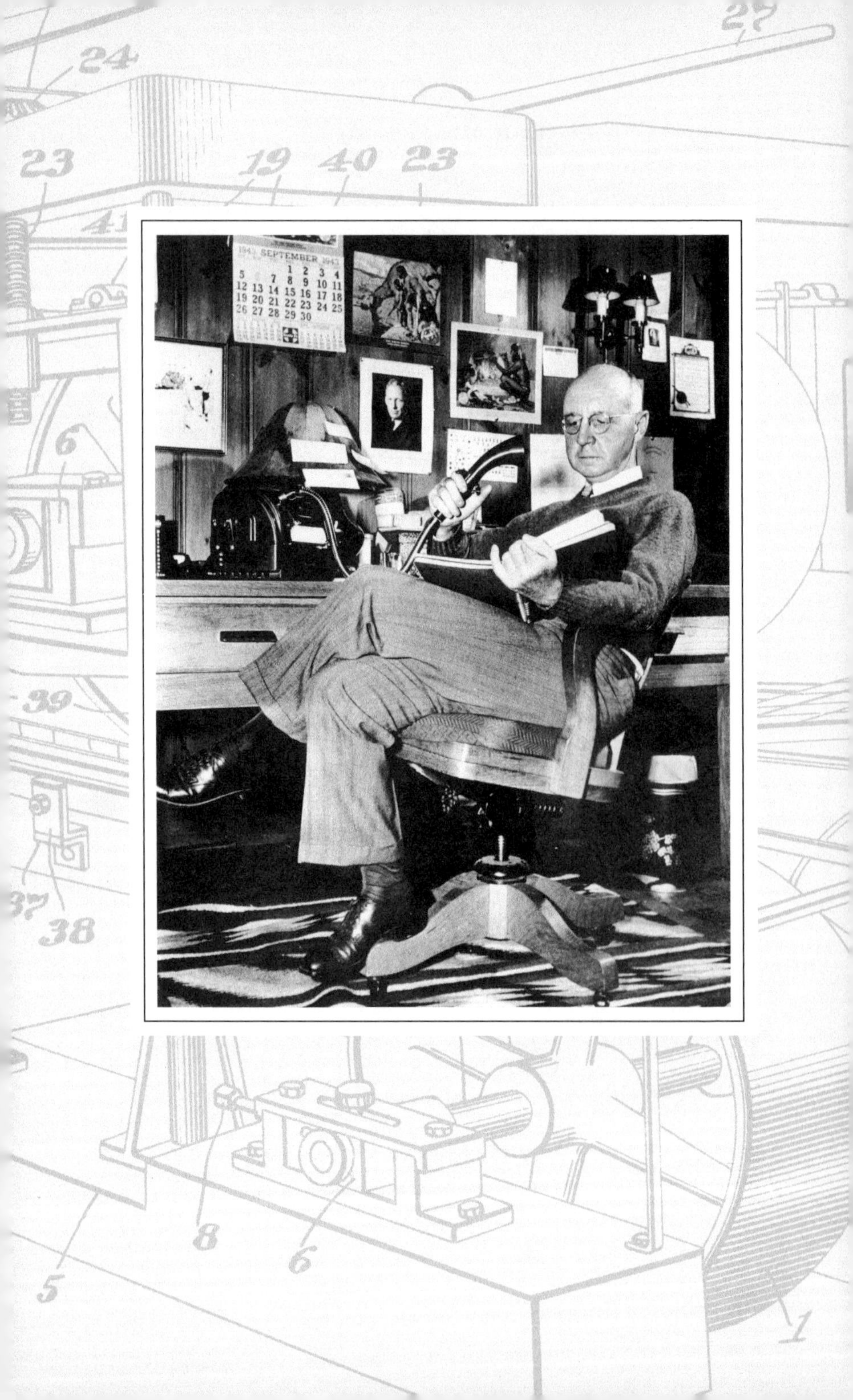

SEPTEMBER 1943

BIRDSEYE

The Adventures of a Curious Man

Mark Kurlansky

Doubleday

New York   London   Toronto

Sydney   Auckland

DOUBLEDAY and the portrayal of an anchor with a dolphin are registered trademarks of Random House, Inc.

Pages 239–40 constitute an extension to this copyright page.

Book design by Michael Collica
Jacket design by Michael J. Windsor
Jacket photograph courtesy of Pinnacle Foods. Birds Eye® is a
registered trademark of Pinnacle Foods LLC.

Library of Congress Cataloging-in-Publication Data
Kurlansky, Mark.
Birdseye : the adventures of a curious man / Mark Kurlansky.—1st ed.
p. cm.
Includes bibliographical references and index.
1. Birdseye, Clarence, 1886–1956. 2. Frozen foods industry—United States—History. 3. Inventors—United States—Biography. 4. Businessmen—United States—Biography. I. Title.
HD9217.U52B575 2012
338.7'66402853092—dc23
[B]
2011044891

ISBN 978-0-385-52705-7

MANUFACTURED IN THE UNITED STATES OF AMERICA

1 3 5 7 9 10 8 6 4 2

First Edition

To the memory of Linda Perney.
The best of friends, who always listened and laughed
but left too soon.

To be perfectly honest, I am best described as just a guy with a very large bump of curiosity and a gambling instinct.

—CLARENCE BIRDSEYE, *The American Magazine*,
February 1951

Curiosity is one of the permanent and certain characteristics of a vigorous intellect.

—SAMUEL JOHNSON, *The Rambler*,
March 12, 1751

Contents

Preface

I never met Clarence Birdseye, who died shortly before I turned eight years old in 1956. I never met any of his seven siblings nor his wife, Eleanor. I met only one of his four children, the eldest, Kellogg, whom I interviewed when researching my book about cod. His grandchildren were all very young when Birdseye died and know little of him. I did interview two surviving sisters-in-law and numerous people who had known him.

The work of a biographer often seems to resemble that of a detective, chasing down clues. Birdseye left a scattered and incomplete record. There is no autobiography or memoir. Surprisingly, the only book Birdseye has left us is a small volume on gardening, which was mostly written by his wife. Huge help was provided by one of the grandsons, Henry, who had more than twenty unopened boxes, which, once he was easily persuaded to open them, revealed a few treasures, including numerous letters from Labrador and Peru. Another grandson, Michael, had a few pieces of the puzzle, such as a collection of his patents and some lost magazine articles. Kellogg, the son, had eight leather-bound handwritten journals from Birdseye's time in Labrador, which Kellogg's widow, Gypsy, donated to Amherst College. They show a great deal about his quirks and interests as well as activities but were not laid out in a way to give a later historian a clear picture of events.

Birdseye's colleagues and collaborators wrote about him. Birdseye wrote numerous articles about himself and his ideas. But the subject himself is not always an infallible source either, especially a man like Birdseye who had an image of himself that he wanted to promote—a very American image of a lot of audacity, not much intellect, and a pioneer spirit. The first and last of these were largely true, the second manifestly not. There are many brief articles on the life of Clarence Birdseye, some written in his lifetime, some at the time of his death, some later. Most of these articles say the exact same things—some true and some not, often with incorrect dates. Unfortunately, these articles appear to have been the source for more articles, and so often the errors are repeated. There are at least forty encyclopedias and dictionaries that contain entries on Clarence Birdseye ranging from general, to business, to inventors, to food. But these too are full of errors and conflicting information.

When I decided to write this book, I had already written about Birdseye in three of my books—the history of cod, the history of salt, and the history of Gloucester. It was time to separate myth from truth and find out who this man with the funny name who changed our way of life really was.

Undeniably, Birdseye changed our civilization. He created an industry by modernizing the process of food preservation and in so doing nationalized and then internationalized food distribution. Birdseye was among the first to talk about that internationalization. In a speech at Montreal's McGill University in 1943, he predicted the postwar world when he announced, "Tomorrow the industry will become truly international." And he was right. This facilitated urban living and helped to take people away from the farms, so that by the early years of the twenty-first century, for the first time in human history, the majority of people on earth lived in an urban area. This would have been no surprise to Birdseye, who often spoke of how improved food preservation made urban life possible.

For many people a major issue is the internationalization of food distribution. Birdseye greatly contributed to the development of industrial-scale agriculture. He even worked with farmers to make their products more suitable for industry. But unlike people today who have grown distrustful of big business, for Birdseye, a product of the zenith of the Industrial Revolution, "industry" was always a good word, without negative connotations. Today's locavore movement—the movement to shun food from afar and eat what is produced locally—would have perplexed him. Why, Birdseye would have wondered, would you want to be limited by local production when the food of the world is available? What would he have thought to see that in his hometown of Brooklyn and his adopted home of Gloucester, there are open-air markets selling local produce when consumers could go to a supermarket and buy the food of California, France, and China for less money? It would have made little sense to Birdseye to prefer small artisanal farms with low and inconsistent yields to the miracles of agribusiness.

Birdseye loved food, loved to cook, and wrote, thought, and talked about eating much more than most people. He was what would be called today a foodie. But he was a nineteenth-century foodie, a foodie in reverse who ate wild local food and artisanally made products, the food of family farms, but who dreamed of making food industrial. It can and probably always will be debated whether this is a good thing. The real argument is whether in changing our eating habits, Birdseye made life better. He personally had no doubt that what he was doing was improving life, but that is largely because he considered his frozen food to be fresh. The idea still exists in the odd commercial oxymoron "fresh frozen." An increasing number of people today ask what this phrase means. How can it be fresh if it is frozen? The claim is that it was preserved in a fresh state. When Birdseye was developing fast freezing, the product seemed remarkably fresh when compared with the existing preserved foods—slow frozen, canned, salted, or smoked.

My grandfather was a tailor, and he hand made all of my father's clothes. And so my father grew up dreaming of someday owning a factory-made suit. We are all more reactive against the conditions we inherit than we realize. In the postindustrial world we have become anti-industry, and it is useful and fascinating to get to know a man of vision and imagination who genuinely believed in industrial answers to life's problems. To understand Birdseye in the context of his times, we need to grasp that people who are accustomed only to artisanal goods long for the industrial. It is only when the usual product is industrial that the artisanal is longed for. That is why artisanal food, the dream of the food of family farms, caught on so powerfully in California, one of the early strongholds of agribusiness with little tradition of small family farms. Birdseye came from a world that was becoming rapidly industrialized, and yet one in which food production was lagging behind and still mostly artisanal.

Birdseye was a naturalist, fascinated by every mammal, reptile, insect, and plant he ran across. But he was not an environmentalist and did not think about the effects of urbanization, industry, and agribusiness on habitat. He was interested in increasing the catches of fishermen, not in the issue of overfishing—a concept that seems never to have occurred to him.

It seems certain that were he alive today, he would see things very differently—and would turn his inventive mind to solving today's problems. Birdseye was a man who kept up with new developments and always wanted to know the latest.

Solving problems was his primary interest. He was a rare and original man who lived in the forefront of his times, operating through imaginative thinking, skilled hands, and great daring. He was a wonder then, and he would be a wonder today.

BIRDSEYE

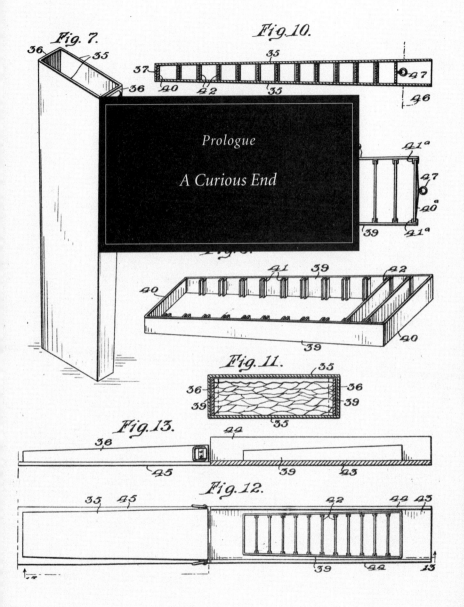

Prologue

A Curious End

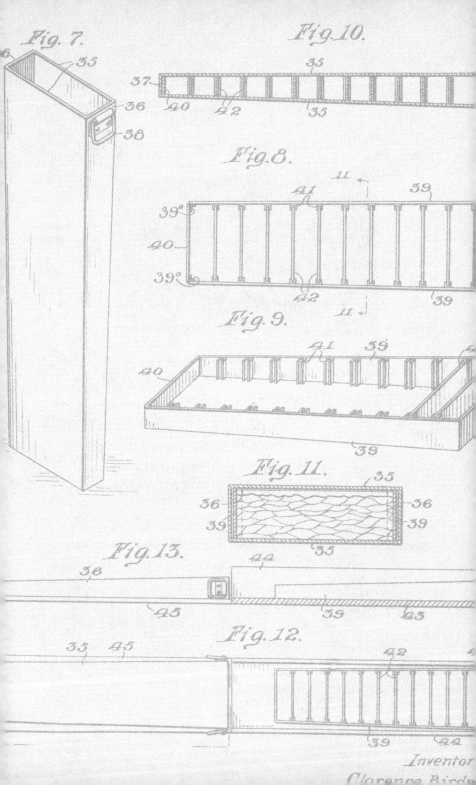

Fig. 7.

Fig. 10.

Fig. 8.

Fig. 9.

Fig. 11.

Fig. 13.

Fig. 12.

Inventor
Clarence Birds

The lobster boats made in Deer Isle, Maine, in the 1940s and 1950s were cumbersome vessels with rounded wooden hulls that made them roll in the slightest swell. Even on a flat sea they were hard to direct, and a steady course could only be achieved by spinning the wheel far to starboard, then, when the bow started to shift, spinning it back far to port. If it was done right, the result could be a straight line, but this was not easy to accomplish. Heading into the wind was impossible, so the boat had to be steered at angles like a tacking sailboat.

In 1956, Sarah Robbins had just bought such a boat and kept it by her home in the old fishing port of Gloucester, Massachusetts. She had not in the least mastered the stubby thirty-five-foot craft when she offered to take her friend Clarence Birdseye on his last adventure. Though Birdseye was more than a generation older, their friendship had been inevitable. They both lived on the fog-swept opening of Gloucester Harbor, an area called Eastern Point, and they were both self-taught naturalists. She had bought the lobster boat to aid her in bird-watching, an interest she shared with Birdseye.

Birdseye, a tiny man, smaller than many of the kids in the neighborhood, with a bland, gray appearance and the dull nickname Bob, was a source of endless fascination in affluent Eastern Point. It was not just that he was famous. Or that he had lived a life of adventure and was full of stories about the

Rocky Mountains and the Southwest at the beginning of the twentieth century and the wild frozen frontier of Labrador before World War I. It was that he seemed to be interested in almost everything and knew a great deal about most of it.

"If I see a man skinning a fish, for example," he wrote about himself in 1951 in *The American Magazine*, "a host of questions pop into my mind. Why is he skinning the fish? Why is he doing it by hand? Is the skin good for anything? If I am in a restaurant and get biscuits, which I like, I ask the chef how he made them: What did he put in the dough? How did he mix it? How long did the biscuits bake? At what temperature? When I visit a strange city, I go through the local industrial plants to see how they make things. I don't care what the product is. I am just as much interested in the manufacture of chewing gum as of steel."

Birdseye died with more than two hundred patents to his name on more than fifty ideas, and though the obituaries called him "the father of frozen food," his inventions ranged from a whaling harpoon to electric lightbulbs. A few of his inventions changed the course of the twentieth century. But it is almost as telling to know that when this enthusiastic and insatiably curious man died, his final mourners were not the captains of finance and industry with whom he worked, nor fellow inventors and thinkers, but the children who grew up in his neighborhood.

When Josephine Swift Boyer was a child in California, her family spent summers at their home on Eastern Point. Like a number of girls in the neighborhood, she took up an interest in birds. When she found a dead rail, a rare specimen for Gloucester, she wanted to keep it, and not knowing what to do, she went to the curious Mr. Birdseye. Not surprisingly, he turned out to be a skilled and experienced taxidermist. He had the right tools and chemicals, and he showed her how to open it and rub it with preservatives and stuff it. The girls, now older women, still remember what delicate and skillful hands he had. Josephine's sister, Lila, said that her clearest memory of him was his fine hands. A reporter from the *New York Post*, interviewing him in 1945, wrote that he had "powerful hands," which is a surprising feature for such a small man.

A family man with four children, he liked helping and talking to young people. "He was very pleasant and easy for a youngster to talk to," said Nancy Ellis, who also knew him as a girl growing up in Eastern Point.

"He was a character," Lila said. "You couldn't help being fascinated by him." He was always building strange things in his large, high-ceilinged basement, or in the kitchen, or even on the lawn of the stately seventeen-room mansion he had built. He was a hunter and was known as a man who would shoot, freeze, dehydrate, or just eat almost anything. Lila, who loved birds and, even in her old age, nursed wounded ones back to health, recalled a strange contraption on Birdseye's lawn. "It was for capturing starlings. That elegant house, and he sat by the pillars ready to pull a string and catch the starlings. We always thought he was going to eat them, probably fast frozen, taste tested for some experiment. In any case we knew they were goners."

Birdseye's wife, Eleanor, was more reserved. But she had been with him for more than forty years, from Labrador to Peru. She too had great curiosity and boundless enthusiasm and a sense of adventure that she probably inherited from her father, one of the founders of the National Geographic Society. Dotty Brown, an Eastern Point girl who knew the family well because she was a close friend of Henry, one of the Birdseye sons, said of Clarence and Eleanor, "They were a devoted couple . . . They were interested in everything. Everything was grist for his mill. He had to know how things grow and what their name is. He didn't just sit around reading the newspaper. Mrs. Birdseye was a little shy, but Mr. Birdseye was outgoing. You couldn't be shy and find things out. He was always asking questions."

Birdseye once said, "Enthusiasm and hard work are also indispensable ingredients of achievement." Perhaps he had too much of both, for he developed a heart condition and often ignored doctors' advice to ease up. On October 7, 1956, Birdseye died of heart failure in his New York City apartment.

Birdseye had requested that he be cremated and that his ashes be scattered past the breakwater where Gloucester Harbor opened to the North Atlantic. It was for this purpose that Sarah Robbins's clumsy Deer Isle lobster boat had been called into service. Sarah was at the wheel, desperately trying to steer a straight course. Eleanor Birdseye stood formally at the stern with Dotty Brown and Lila Monell. At the bow was the son Henry Birdseye, holding a white ceramic urn with the ashes of his father inside. Sarah managed to stay off the granite blocks of the breakwater and sail out of the harbor opening and hard aport to the open Atlantic, following the route, albeit with a few more curves and zigzags, sailed by hundreds of thousands of Gloucester fishermen since the town became a fishing port in 1623.

Of the four children, Henry was in many ways the most like his father. He too was described around Eastern Point as "a character." He spent a great deal of time at the Eastern Point Yacht Club and like his father loved to amuse people with great stories. They used to call him Birdseed. Also like his father, he had a bent for science and borrowed a corner of the basement laboratory to build his own chemistry lab. Like his father, he had an entrepreneurial side and became an expert scuba diver, operating a salvage business around Gloucester Harbor. During World War II, he was a decorated soldier in the Pacific and later became an avid flier, geologist, and amateur astronomer.

But Birdseed at the bow did something unexpected in front of his mother and old friends. He ceremoniously tossed the white ceramic urn into the gray-green Atlantic without opening it to scatter the ashes. The problem was that it did not sink. It floated and bobbed like a white lobster buoy.

"Break it, break it!" Dotty and Lila shouted hoarsely, trying to get his attention without Mrs. Birdseye noticing from the stern. If the urn was left to float, it would surely wash onto one of the popular beaches along what was called the Back Shore. Who would find it? What would happen to it? Would this be one of those slightly weird local stories in the *Gloucester Daily Times*?

"Break it!" Dotty shouted again, handing him the boat hook while Lila retreated to the stern to keep Mrs. Birdseye occupied. Sarah turned her wheel wildly, trying to bring the boat about to give Henry a shot at the urn. He poked at it, but it didn't break, and she had to make another pass. Turning circles in a boat that will not head into the wind is no small task. But the urn would not break. Meanwhile, Lila was chatting with Mrs. Birdseye, hoping she did not notice her son jabbing at the sea. Finally, Henry was able to get the right angle to crack the urn with a blow from his hook, and the last of Clarence Birdseye drifted into the tide like plankton.

Had Clarence Birdseye been there in the flesh observing

all this, he probably upon landing would have hopped into one of his three cars, driven downtown to a hardware store to buy a few things, and retired to his basement to build either the fast-breaking funeral urn or the quick-sinking one and then would have patented it, certain that in the future millions would end their days in a Birdseye urn. That's who he was, a man who observed how things worked and figured out how to make them work better.

"He always wanted to improve something," said Lila Monell.

When he was born in Brooklyn not quite seventy years earlier, there had been little to suggest what an adventure his life would be. But when he heard people comment on what an improbable adventure his life was, he would shrug and say, "I was just very curious."

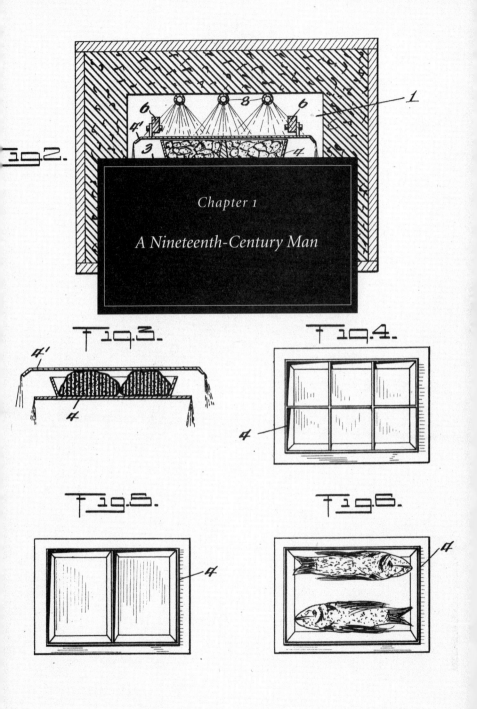

Fig. 2.

Fig. 3.

Fig. 4.

Fig. 5.

Fig. 6.

Chapter 1

A Nineteenth-Century Man

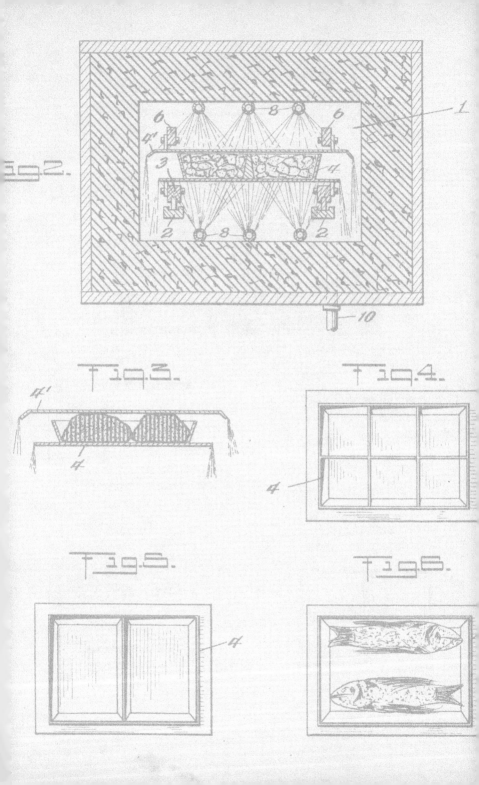

Fig.2.

Fig.3.

Fig.4.

Fig.5.

Fig.6.

Clarence Frank Birdseye II was born in Brooklyn on December 9, 1886. Both the year and the place are significant. In 1886, Brooklyn was a separate city from Manhattan and, in fact, was the third-largest city in America and one of the fastest growing. Between 1880 and 1890 the population grew by more than a third to 806,343 people.

One of the forces that made this dramatic growth possible in Brooklyn and neighboring Manhattan was refrigeration. Because of this new technology a large population could live in an area that produced no food but rather brought it in and stored it. Natural ice, collected in large blocks from the frozen lakes of New England and upstate New York, was stored in sawdust-insulated icehouses built along the Hudson that shipped all year long. New York City used more than one million tons of natural ice every year for food and drink. While the pleasure of iced drinks in the summer had been a luxury of the wealthy ever since Roman times, in New York at the time of Birdseye's birth it had become commonplace. Almost half of all New Yorkers, Manhattanites and Brooklynites, kept food in their homes in iceboxes—insulated boxes chilled by blocks of natural ice. A few even had artificially chilled refrigerators, dangerous, clumsy electric machines with unpredictable motors and leaking fluids.

No place else in the world was using this much ice. Birdseye was born into a world of refrigeration and would find it lacking when he left the New York City area. It was one of those things that New Yorkers took for granted.

People are mostly formed over their first dozen years; Birdseye, having been born in 1886, was a nineteenth-century man, even though he lived most of his life in the twentieth century. This, of course, was not unusual. For the first half of the twentieth century, people shaped in the nineteenth century dominated most fields. John Kennedy, elected in 1960, was the first twentieth-century U.S. president. Historians have often commented on how historical centuries do not fit neatly between year 1 and year 99, and quite a few have thought the historical nineteenth century to be an unusually long one, lingering well into the twentieth, whereas the twentieth century to some appears to have been a short one, transitioning even before the year 2000 into a new age that would be associated with the twenty-first century.

Clearly, Birdseye was shaped by the nineteenth century. Even as an inventor, he used nineteenth-century industrial technology for nineteenth-century goals, as opposed to someone like his fellow Gloucester inventor John Hays Hammond, who harnessed radio impulses into such devices as remote control and was very much a twentieth-century inventor. Birdseye's inventions, from freezers to lightbulbs, were all mechanical and never electronic. Yet his impact on how people lived in the twentieth century was enormous.

The nineteenth century, the time of the Industrial Revolution, was an age of inventions, and inventors were iconic heroes. Ten years before Birdseye's birth, Alexander Graham Bell had invented the telephone. The following year Thomas Edison invented the phonograph. The year after that, 1878,

Sir Joseph Wilson Swan, a British inventor, patented the first incandescent lightbulb and lit his house with it. The year before Birdseye was born, a German engineer named Karl Benz patented the first automobile that was practical to use, a three-wheeled vehicle powered by an internal combustion engine and fueled by periodically filling a tank with gasoline. The same year another German, Gottlieb Daimler, built the first gas-powered motorcycle. In 1886, the year of Birdseye's birth, Daimler built the first four-wheeled automobile.

Among the important inventors at the time of his birth, George Eastman was to have a profound effect on Birdseye. In 1884 he patented roll film, and in 1888 he produced a light-weight camera using that roll film, the Kodak camera. He and his company, the Eastman Kodak Company, the first major supplier of photographic equipment, gave birth to amateur photography, a passion of the young Clarence Birdseye. His was the first generation to grow up with amateur photography, and he was a pioneer—from his snapshots in the early twentieth century to home movies in the 1920s to color film in the 1930s.

Inventions of all kinds, ingenious mechanical solutions to practical problems, were popping up seemingly every day. In 1884, the synthetic cloth industry began when a French chemist, Louis-Marie-Hilaire Bernigaud, comte de Chardonnet, patented a process to make artificial silk, which decades later became known as rayon. In 1884, Lewis Waterman, a Brooklyn insurance agent, frustrated with the inefficient pens of the day, invented the capillary feeding fountain pen, the first practical alternative to a pen dipped in an inkwell. James Ritty, an Ohio barkeeper, became the first manufacturer of his new invention, cash registers, the same year. In

1885, Hiram Maxim, an American inventor, demonstrated the first machine gun to the British army. In 1886, in addition to Daimler's automobile, Coca-Cola and the first washing machine were invented. Barbed wire, which divided up the open range and changed the character of the American West, and wearable contact lenses were both patented the following year. In 1888, Marvin Stone, an Ohio-born inventor, came out with the first paper drinking straw. In 1889, Joshua Pusey, a cigar-smoking Pennsylvania attorney, invented the matchbook. In 1891, when Birdseye was four years old, Jesse Reno, the son of the Civil War officer after whom the Nevada city was named, invented the escalator. Typical of his generation of inventors, he was part inventor and part entrepreneur. Reno created a sensation in Birdseye's native Brooklyn when he showcased the escalator for two weeks as a ride at the Coney Island amusement park. It was then featured on the Manhattan side of the Brooklyn Bridge, still in itself a sensation since its 1883 opening as the longest suspension bridge in the world, for the first time connecting Brooklyn and Manhattan.

New York City was being reshaped. The Statue of Liberty was installed in the harbor and dedicated the year of Birdseye's birth. Eight years later the city began digging tunnels for subways.

These were exciting times in New York and in America. In the nineteenth century there was a notable difference between the culture of American inventors and that of Europeans. European inventors reveled in the theoretical, sometimes even shunning patents lest they give the impression of harboring lowbrow commercial interests, whereas Americans thought inventions were pointless without practical and commercial applications. There were of course exceptions.

Daimler, who was an engineer and not a scientist, did start producing automobiles, and Chardonnet, who was a scientist and a colleague of Louis Pasteur's, did start a synthetic textile mill. But many Europeans were content in the world of the theoretical, whereas Americans had a Puritan belief that anyone who invented something had a moral obligation to put it to useful service. The press would criticize inventors who failed to do this.

Dr. John Gorrie ran a naval hospital in Apalachicola, Florida, that treated victims of yellow fever and malaria. He invented a primitive form of air-conditioning in the 1840s that produced artificial ice by expanding compressed air. Modern air-conditioning was not developed until sixty years later at the beginning of the twentieth century. Gorrie cooled his hospital and his home with the device, but he was so attacked by religious conservatives for interfering with God's design that he published the ideas behind his invention under a pseudonym. Not realizing who the writer was, the editor of the publication, the *Commercial Advertiser*, criticized the author for having failed to put his ideas into service.

Robert Fulton, the father of the steamboat, was a prototypical American inventor. The steamboat had had a long development, which had nothing to do with Fulton. It was a European idea, as was the steam engine. A Frenchman, Denis Papin, built a steam piston in 1690 and a steamboat in 1704 but failed to attract any interest in the boat. The Scottish engineer James Watt built a greatly improved steam engine at the time of the American Revolution. But although the French, the Germans, the British, and then the Americans built various steam-powered vessels, there was no *commercially* successful steamboat until Robert Fulton. The great inventor Robert

Fulton did not exactly "invent" anything. But he put the right kind of engine in the right kind of vessel and established a commercial run on the right route. Earlier steamboat lines were on less profitable routes or ones with good alternative land transportation. Fulton established his line on the East River in Manhattan running up the Hudson to Albany. It was the first commercially successful steamboat line in part because there was no good land route for hauling freight between these two commercially important centers. Fulton is often erroneously remembered today as the inventor of steamboats—in much the same way that Birdseye is erroneously remembered as the inventor of fast freezing—but the real reason we still know the name Fulton is that he launched an industry by showing that money could be made from steamboats and that it was a commercially important idea. In America an important idea is an idea that makes money.

Europeans did not always like the American attitude. Albert Einstein, essentially a nineteenth-century European scientist and pure theoretician who found himself in twentieth-century America, wrote of this pragmatic side of American thinking, "There is visible in this process of relatively fruitless but heroic endeavors a systematic trend of development, namely, an increasing skepticism concerning every attempt by means of pure thought to learn something about the 'objective world,' about the world of 'things' in contrast to the world of mere 'concepts and ideas.'"

Birdseye grew up in a world in which mere concepts and ideas were not enough. An American inventor solved a problem, formed a company, and, he hoped, earned a fortune. Alexander Graham Bell, a Scot turned Canadian who then came to America, was most known at the time of Birdseye's birth as not only the inventor of the telephone but the founder of the first telephone company, Bell Telephone, in 1877. By the time Birdseye was born, the Bell Telephone Company had placed phones in 150,000 homes and offices.

Ten years before Birdseye was born, Thomas Edison came to the public's attention for selling his telegraph idea to Western Union for a surprising $10,000. Then, in 1880, he created another industry, Edison Lamp Works, which manufactured fifty thousand lightbulbs annually. When Edison developed an idea, which he did with astounding regularity, he built a company, and he usually made money.

Inventors were founders of industries, not intellectuals.

The other great influence of the nineteenth century that would never leave Birdseye was America's westward romance. Nine years before Birdseye's birth the Nez Percé fought the U.S. cavalry in northern Montana in the last great battle of the Indian wars.

During Birdseye's childhood, according to several historians, the most famous person in the world was Buffalo Bill Cody. He had earned his nickname after having been hired by the Kansas Pacific Railroad to keep the rail workers supplied with buffalo meat. According to legend, in eight months between 1867 and 1868 he killed 4,280 buffalo. This kind of killing was another nineteenth-century American reality that strongly influenced Birdseye, who loved to hunt. Historians have labeled the second half of the nineteenth century the age of extermination. The American bison herd, the largest land animal herd ever recorded, was by the late nineteenth century reduced to only a few hundred. Passenger pigeons, which were thought to number 5 billion when Europeans started coming to North America and may have represented as much as 40 percent of the continent's bird population, were exterminated for cheap food. The last one died in a Cincinnati zoo in the early twentieth century.

In the nineteenth century 400,000 skunks, 500,000 raccoon, and 2 million muskrat were killed in a typical year. Beavers, seals, and sea otters were all victims of the fur trade. Gradually influenced by such writers as Ralph Waldo Emerson, Henry David Thoreau, and John Muir, Americans started to grow concerned about the slaughter, and government began regulating hunting. At several points in his youth Birdseye came into conflict with the new hunting regulations.

Born in the Iowa Territory in 1846, William Cody was famous for being a colorful character. He used to boast that he had been a Civil War veteran, a trapper, a bullwhacker, a "Fifty-Niner" in the Colorado gold rush, a Pony Express rider in 1860, a wagon master, and a stagecoach driver. Most of it was true, and oddly it was not that different from the way Clarence Birdseye, who also wanted to be a colorful character, would years later describe himself. In 1951, Birdseye wrote, "The public customarily thinks of me as an inventor . . . but inventing is only one of my lines. I am also a bank director, a president of companies, a fisherman, an author, an engineer, a cook, a naturalist, and a dock-waloper." Some of these claims are plainly true. Fisherman, author, and cook are exaggerations, and it is not even clear what he is referring to when he claims to have been a "dock-waloper."

Cody was the late-nineteenth-century prototype of a "colorful character," and he traveled the world with his show of taming the West. Included in his cast was Annie Oakley, Sitting Bull, who had inspired the most famous Indian victory over the U.S. Army in 1876 at the Little Bighorn River in the Montana Territory, and numerous other formerly hostile Indians whom he paid well to act out their defeat. When Birdseye was a child in Brooklyn, Buffalo Bill was living nearby, in Staten Island.

What is the relevance of the Wild West, the age of extermination, and Buffalo Bill Cody to a processed-food inventor from Brooklyn? Very early in Birdseye's life, the connection became evident.

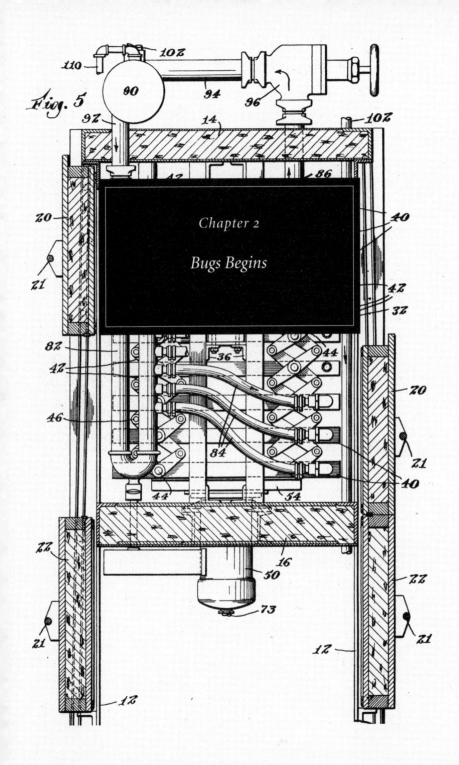

Fig. 5

Chapter 2

Bugs Begins

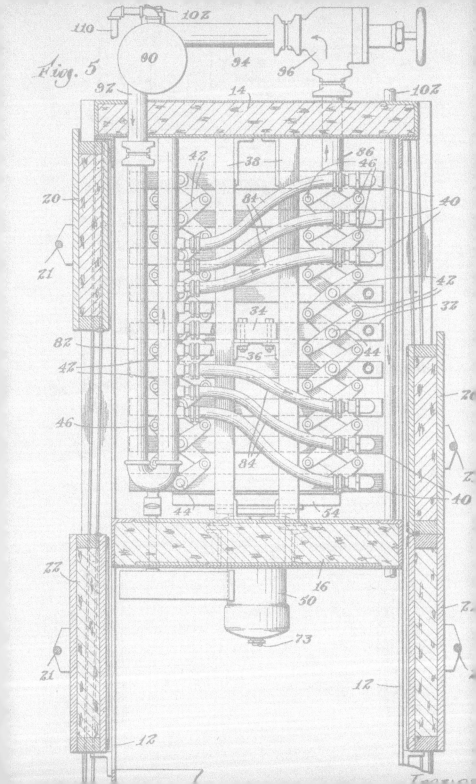

Fig. 5

As an adult, Clarence Birdseye loved to tell stories, and one of his favorite tales was about the origin of the name Birdseye. According to him, it was originally two words—Bird's Eye. In his telling, the young page to an English queen saw a large hawk swoop down toward Her Majesty. At this point, Birdseye would explain, "this page boy ancestor of mine, according to the records, took out his trusty bow and arrow and shot that bird right in the eye. The queen was so tickled she gave him the name right on the spot." And he added, she also gave him the family motto, which is "Stay right on the target."

He never produced these records he referred to nor the name of the queen or even explained exactly in which century the incident was supposed to have taken place. The hunting skill, making the improbable shot, was so typical of Birdseye's kind of story that there is reason for suspicion. But it did seem an odd name, and all his life people had trouble believing that Birdseye was really his name. Another one of his favorite stories was how he found himself in San Francisco and short on money. Wanting to cash a check, he called the local branch of the Birds Eye Division of General Foods, a company named after him. He explained his problem and gave his name, and the man on the other end of the line told him if he stopped kidding and gave his real name, he would connect him with someone. Even today, while the brand name Birds Eye is well known, many people are surprised to learn that there really was someone named Birdseye.

According to the official family history, the American Birdseye story begins with a Puritan from Berkshire, England, named John Birdseye who settled in New Haven, Connecticut, in 1636. This is probably off by a year because Puritans from England settled New Haven in 1637, although 1636 could be right if, as other records suggest, he settled in Wethersfield. A son, also named John Birdseye, was among the founders of Milford, Connecticut. A grandson, the Reverend Nathan Birdseye, lived on a country estate in Stratford until his death in 1818 at the age of 103. Though he lost both his sight and his hearing in old age, he continued preaching almost up to his death and was known as a charming storyteller. He had twelve children and claimed 206 living relatives at the time of his death, ensuring the presence of a considerable number of Birdseyes in America ever since. Nathan's eldest son, Captain Joseph Birdseye, was an officer fighting the British in the American Revolution. Victory Birdseye, born in 1782, became a prominent lawyer in upstate New York, beginning a family tradition in law. He was a state representative, a state senator, and a U.S. congressman. His son Lucien became a prominent Wall Street attorney and settled in Brooklyn. When he died in 1896, the *New York Times* called him "one of the best known lawyers in this city and Brooklyn."

Lucien's son Clarence Frank Birdseye, born in 1854, went to an elite Brooklyn school, graduated from Amherst in 1874 and Columbia Law School in 1877, and, like his father and grandfather, became a prominent attorney. He authored numerous books, often on education, such as *The Reorganization of Our Colleges* and *Individual Training in Our Colleges*, as well as a number of legal textbooks, including his 1896 book of New York State code and statutory law, which became a standard legal reference book. He also wrote *American Democracy Versus Prussian Marxism*, which is largely a diatribe against socialism in which he also reveals a deep dislike for all things Prussian. He wrote the book after World War I, when Prussianphobia was in vogue. Birdseye's thesis was that Marx was born in Prussia, went to university in Prussia, and collaborated with Engels, a fellow Prussian, and all his thinking was rooted in "Prussian ruthlessness."

Clarence married Ada Jane Underwood, who was born in Brooklyn in 1855. The Underwoods moved to Tolland, Connecticut, where Clarence and Ada were married in 1878. Sixteen years later, Clarence's older brother Henry Ebenezer, a Wall Street financier, married Ada's younger sister Annie. That marriage, in 1894, also took place in Tolland.

If the urge toward industrial invention has a genetic basis, the younger Clarence inherited it not from the legalistic Birdseyes but from his mother's side. Ada's father, Henry Underwood, was a belt manufacturer. The machines of the Industrial Revolution ran on belts that connected motors to working parts through pulleys. Henry Underwood reasoned that the belts would stop slipping if they had more surfaces on the pulleys. So rather than a flat or rounded belt, he built a trapezoidal one to run in pulleys with angular grooves. It did not have the impact of electric lightbulbs or even frozen food, but it improved many machines and merited mention in *Scientific American* in its roundup of new inventions in April 1860, when Underwood patented his idea. Underwood was the first American manufacturer of leather belts and later improved on them with cotton-leather belts. In the 1890s,

Underwood belts and those inspired by his inventions were critical in the growth of industrial machine technology and, ironically, were a small but essential part of the development of refrigeration machines.

Clarence and Ada lived in Brooklyn, at 368 Clinton Street, in a handsome corner brick-and-brownstone four-story house on a tree-lined street where such fine homes were built in rows. They had a long brownstone stoop and a wrought-iron rail with ornate masonry above a decorative carved wooden doorway with panels of frosted etched glass. A corner building was particularly desirable because it meant that the side of the building and not just the front had windows and light. In fact, the side on Sackett Street had ornate carved wooden bay windows. The rooms had high ceilings and tall arched doorways and chandeliers hanging from the ceilings. The neighborhood, Cobble Hill, was one of the oldest in Brooklyn, first settled by the seventeenth-century Dutch, who called it Punkiesberg, but it got a new life in the nineteenth century with a Brooklyn building boom for a growing upper-middle class—people like the Birdseyes. They started building houses by the hundreds, first in affluent Brooklyn Heights and then in the areas around it, such as Cobble Hill.

Clarence and Ada had their first child, Miriam, in 1878 and their second, Kellogg, in 1880. Their third, Henry, was born in Tolland, Connecticut, though there is no record of why they were there, and then the rest of their children were born in Brooklyn. They had nine children, and one of the twin girls born two years before Clarence, Marjorie, died at the age of not quite nine months. Clarence was their sixth child. By the time he was born, their Cobble Hill neighborhood was even more fashionable because the nearby newly completed Brooklyn Bridge brought them closer to Manhattan.

There is not a great deal known about Clarence's relationships with his seven siblings. In his letters and journals they are only occasionally mentioned, but he did name his two sons after two of his brothers—Kellogg and Henry. Both older brothers became businessmen whom he often turned to for advice, and he seems to have been particularly fond of his brother Roger Williams Birdseye, born four years after him. The Birdseyes were a large family in which relatives were available for advice and connections. In the forty years that the restless Clarence rummaged through opportunities looking for his career, relatives constantly aided him.

Young Clarence showed few signs, or at least few have been preserved, of being the garrulous social person he became as an adult. As a child, he disliked organized sports and the other activities that were engaging most boys. It is not known if he had close boyhood friends. Most descriptions of the boy Clarence have him alone in nature. When he was about eight years old, the family bought a farm in Orient on the end of the North Fork of Long Island. The farm was named Wyndiecote, and Birdseye so loved it that he would use the name repeatedly for his homes as an adult. Wyndiecote was his ideal home. "I liked nothing better," Birdseye recalled more than a half century later, "than to tramp alone through fields or along the seashore studying the birds and other wildlife which I encountered." His mother, Ada, was the first to notice that the boy was at heart a naturalist. "I guess I was some kind of naturalist from the time I could walk," Birdseye said and then added, "At least my mother thought so."

When Clarence was ten years old, he became fixated in that way ten-year-olds do on getting his own shotgun. The idea was already entrenched in him that hunting was an essential part of enjoying nature. But it was also a way of turning a profit from nature. Making a profit was always a fundamental idea for Clarence. In his walks through the marshes around the sprawling Birdseye farm, Clarence had noticed a great number of muskrat. In a thought process that would be constantly repeated throughout his life, he wondered where there might be a market for live muskrat. He wrote to Dr. William T. Hornaday, the director of the Bronx Zoo, and asked him if he would be interested in acquiring some muskrat. Hornaday wrote back, explaining that he already had all the muskrat he could use, but he referred him to an English aristocrat who was stocking an estate. The ten-year-old went into the marshes and set traps until he had twelve live muskrat, and he shipped them to England. Nine of them survived the trip, and the Englishman paid him $1 for each. With the $9 he bought a single-gauge shotgun.

He then used the shotgun and learned how to preserve and stuff his victims by reading books and asking questions at taxidermy shops. The following year, in the winter of 1897, when he had just turned eleven years old, he placed an ad in a sports magazine for "the American School of Taxidermy." In reality Clarence was all there was to the American School of Taxidermy. But he was offering courses at modest rates. It is easy to imagine the comic scene unfolding when someone wishing to learn the secrets of taxidermy answered the ad and discovered that the entire school consisted of an eleven-year-old boy. But unfortunately, there is no record of anyone responding.

Some of those who have told Birdseye's story in magazines and newspapers have suggested that he was doing these kinds of things because the family finances had fallen on hard times. This was in fact to happen some years later, but at this point the Birdseyes seemed to still be living well. Like a good Puritan inventor, Birdseye simply liked to see a good idea translated into a commercial success.

There was also something else moving him. He longed for a different kind of life, a life of adventure. He was fascinated not only by Buffalo Bill but by all kinds of tales of the American West, of cowboys, and hunters, and trappers. His favorite writer was George Alfred "G. A." Henty, a nineteenth-century English war correspondent turned popular novelist who wrote adventure stories, usually for boys. The lead characters were always highly intelligent, extremely resourceful, and courageous. He wrote 122 novels before he died in 1902. Birdseye's favorite book by Henty was his 1891 novel *Redskin and Cowboy*. It is the tale of an English boy who is forced to leave the safety and quiet of his uncle's home and journey to the American West, where he meets up with such characters as Straight Charley, Broncho Harry, and Lightning Hugh. Although most Henty novels are thought of as children's books, Birdseye read them, as well as other westerns, his entire life and especially continued to reread *Redskin and Cowboy*. In a 1945 interview he told the *New York Post* that *Redskin and Cowboy* was the book that "first influenced him to live the outdoor life."

When Clarence was entering high school, the family moved from their Cobble Hill brownstone to Montclair, New Jersey. Montclair was a destination for successful New Yorkers, not

New Yorkers down on their luck. Montclair became independent of Newark because the citizens wanted better rail service than the city was providing. In the last third of the nineteenth century there was a sizable immigration of affluent New Yorkers to the new town. Businessmen and also artists came and built ample homes. Montclair had even had a recognized painting movement in the 1870s centered on George Inness, a Newark native who settled there in 1885 and was celebrated for the saturated color and brilliant light of his Montclair landscapes.

In high school, although his interests were primarily in science, Clarence also took the unusual step of enrolling in a cooking class. This was at a time when progressive women who wanted to teach workingwomen offered cooking classes, teaching them how to prepare food quickly while still maintaining jobs. The idea grew out of a mid-nineteenth-century movement to educate women on home management while men were being educated on matters outside the home. It included cooking and the science of nutrition and would eventually be called home economics. Catharine Beecher, sister of Harriet Beecher Stowe, was one of the early proponents who fused modern science with home care. Ellen Swallow Richards, the first woman to graduate from MIT, a chemist known for her research on the quality of drinking water, was another pioneer in applying science to domestic studies. Fannie Merritt Farmer, who graduated from the Boston Cooking School in 1889, greatly popularized the "scientific" approach to cooking, incorporating notions of simplicity, frugality, nutrition, and sanitation. Juliet Corson was a Bostonian who established a cooking school for poor women in New York, teaching frugality and proper nutrition from her home on

St. Mark's Place. Shortly before Birdseye's birth she was teaching one thousand students a year, charging a nickel to poor students and extravagant fees to the wealthy.

By the time young Birdseye was in high school in Montclair, frugal, scientific, pragmatic, nutritious cooking was the established approach. His older sister Miriam, who graduated from Smith in 1901 and then earned a graduate degree in domestic science from Pratt in 1907, may also have influenced Clarence. She started as a home economics teacher in New York City and later became a prominent nutritionist.

It is unfortunate that little is known about life in Ada and Clarence senior's household that would lead two of their children toward distinguished careers in food. Another sister, Katherine, the surviving twin two years older than Clarence, would later marry John Lang, divorce him, settle in Atlantic City, New Jersey, and become president and general manager of Mrs. Lang's Candy Kitchen, which sold candy in stores along the boardwalk.

After graduating from high school, Clarence took a job as an inspector for the New York City Sanitation Department and then a job as a $3-a-week office boy in a Wall Street financial firm. According to legend, he was on a lunch break when he found on the ground part one of a ten-part mail-order shorthand course. He studied his part one and from that developed his own form of shorthand, which he used most of his life. Exactly what the system was, how to read it, and what he used it for remain murky. When his son Kellogg was first married, Birdseye tried to persuade his new daughter-in-law, Gypsy, to transcribe some manuscripts from his shorthand. But he gave her very little instruction and she quickly gave up the project in frustration.

Clarence worked through the summer, but in the fall his parents sent him off to elite Amherst College, in Massachusetts, at age nineteen, where they supported him in the style of a wealthy son. There was still no sign of family financial difficulties. Student rooms were available for as little as $55 a year, but the Birdseyes got Clarence the top-priced room at $120 a year, a steep price for a boy who had been working on Wall Street for $3 per week.

Amherst was the Birdseye family school, in the rolling green foothills of the Berkshires, the hometown of the poet Emily Dickinson, who died the year Birdseye was born and was just becoming famous while Birdseye was in college. This was the kind of thing Amherst was known for—its literature program. It was a liberal arts college with no emphasis on science. Science majors such as Birdseye were awarded bachelor of arts degrees just like literature majors. There was no bachelor of science degree. So it was not the ideal college for young Clarence, who cared, above all else, about science. But not only had Clarence's father graduated from Amherst, so too had his older brother Kellogg, in the class of 1902, and another brother, Henry, attended but didn't graduate, which would also be true of Clarence's younger brother, Roger.

At Amherst, Birdseye studied under respected nineteenth-century men, many of them with the long and elaborate hair and beards of that earlier century. The college president, George Harris, had graduated from Amherst with the class of 1866. The professor in Birdseye's favorite subject, biology, was John Mason Tyler, who had graduated from Amherst in 1873, back when the writings of Charles Darwin were fresh off the press and not yet part of the accepted teachings in the field. Benjamin Kendall Emerson, with whom Birdseye hoped to study geology, graduated from Amherst in 1865, the year the Civil War ended. Edward Hitchcock, a doctor who taught physiology and anatomy and was a pioneer in the physical education program with which Birdseye struggled, was born in the town of Amherst in 1828, the year the first passenger railroad in America started construction and Noah Webster's dictionary of the American language was first published.

In Clarence's first year he was an honors student in science but only average in everything else, including physical education. He did particularly badly in Spanish. Clarence's fellow students called him Bugs because he was forever examining some bug or rodent in the nearby countryside. It is not known if Bugs grew out of Bob or the other way around, but after about 1906 almost no one called him Clarence.

Birdseye the naturalist never stopped working. Behind the town butcher shop was an infestation of rodents—what would have been horrible black rats to most people. But Birdseye recognized this particular rat as a nearly extinct species, *Mus rattus*, and was able to find a geneticist at Columbia University, T. H. Morgan, who wanted them to crossbreed with another wild species for experiments to better understand why in heredity some features are dominant and others recessive. Morgan was willing to pay $135 for a shipment of live *Mus rattus*, which was more than the $110 annual tuition at Amherst.

Birdseye wrote in later accounts that he used to spend free time at Amherst wandering the fields with his shotgun on his shoulder. "Suddenly I came upon an open spring-hole where thousands of small frogs were congregating—layers and layers of them came to hibernate for the winter.

" 'What are those frogs good for?' I asked myself."

It was seven years later, but he thought he would try the Bronx Zoo again since it had helped him sell the muskrat. He knew that the zoo had frog-eating reptiles that needed to be fed. It did want them, and he shipped them live—reptiles only eat them live—wrapped in wet burlap. For this he earned $115.

He seemed to feel isolated at Amherst and did not appreciate being called Bugs. He avoided talking about the frogs, the rats, and other activities because they led to ridicule. Writing about his college days years later, he said, "I did not realize at the time, as I have discovered since, that anyone who attempts any thing original in this world must expect a bit of ridicule."

The summer between his freshman and his sophomore years he went to New Mexico for the U.S. Department of Agriculture. In 1908, for reasons that are not clear, the Birdseyes fell into financial crisis. It seemed to happen just as Clarence was finding his rhythm in school. He was not athletic and played no sport and did not do well in gym. But he did join a fraternity, something in which his father was a great believer and about which he had even written a book. By the spring of 1908 young Bugs was scoring as high as 96 percent in biology and had almost a 90 percent overall course average, which was extremely rare. He applied to take special advanced biology and geology classes.

But there was no more money for schooling. On December 31, 1908, he wrote to the school registrar, "It will almost certainly be necessary for me to go through the rest of this year without any financial aid from home; and to this I must borrow about $300—$150 or $200 at once. Is there a student loan fund at Amherst and if there is can I secure the necessary money at not more than six percent interest and for one or two years?"

Apparently, he was not given a loan, because after the spring of 1908 he dropped out of college. This was the end of his formal education.

What was he to do with only two years of college? Without his college degree he could not become the fourth generation of distinguished attorneys. But he was not likely to have done that anyway. He had wanted to go to school to learn, not to launch a career. He could have returned to Wall Street and tried to make his mark in business, as New York was full of possibilities. But that was not what Birdseye was inclined to do. When he was in his sixties, he said, "Any youth who makes security his goal shackles himself at the very start of life's race." Birdseye was looking for adventure. And where did you find adventure in 1908? In the West.

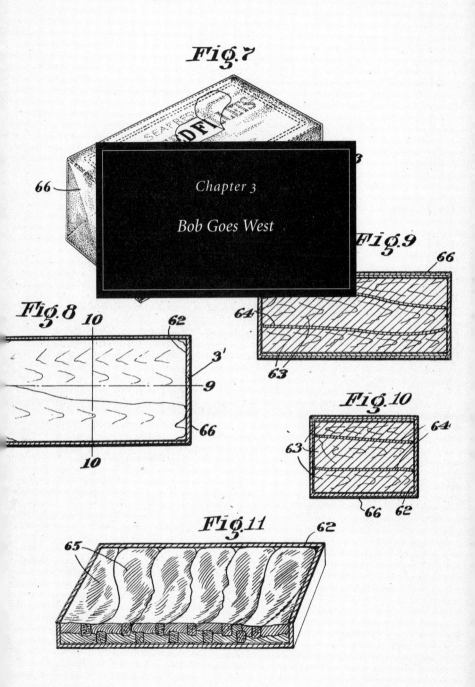

Fig.7

Fig.8

Fig.9

Fig.10

Fig.11

66

10

62

3'

9

66

10

64

63

66

63

64

66

62

65

62

Chapter 3

Bob Goes West

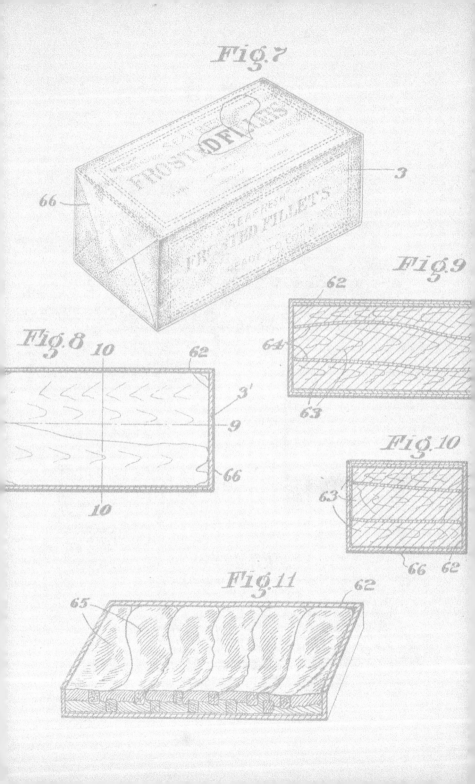

Fig. 7

66

3

Fig. 8

10

62

3'

9

66

10

Fig. 9

62

64

63

Fig. 10

63

66   62

Fig. 11

65

62

The record is not clear on how Clarence Birdseye decided on Arizona and New Mexico or when exactly he went to work for the U.S. Biological Survey. In his letter to Amherst asking for a scholarship, he stated that he had spent the previous summer, 1907, in New Mexico. After dropping out of college in 1908, he returned. And years later he stated that he worked for the U.S. Biological Survey in the Southwest in the summers of 1908 and 1909. He needed to earn money, and he far preferred to earn it far away in the outdoors than on Wall Street in his native town.

Arizona and New Mexico were still frontier outposts in 1908 with a combined population of fewer than half a million people. They were both territories, ruled by a presidential appointee, and had not yet established the necessary infrastructure to apply for statehood, though they would both soon do this and become states only four years later.

The territories had become increasingly present in the popular imagination, not only from western legends, stories, and novels, but from a landscape painting movement featuring such sights as the Grand Canyon by artists like William Henry Holmes (1846–1933). Holmes was not only a legendary painter and mountain climber in the West but an art curator in Washington, D.C., who did much to popularize pottery, textiles, and other Indian crafts from the Southwest.

Railroads were expanding their reach, though horses were still the primary transportation in the hinterlands. In 1901 the railroad reached the Grand Canyon, and the Arizona tourism industry was founded with a tourist village on the southern rim. The Fred Harvey Company, America's first chain restaurant, which followed the railroad throughout the West, established a restaurant on the southern rim in 1905.

The U.S. Biological Survey, founded in 1885, was a branch of the Department of Agriculture largely engaged in the wholesale extermination of wild animals considered a nuisance by farmers and ranchers. Its leading victims were wolves and coyotes, and it hired hunters and trappers to kill them. In 1907 alone the program killed 544 coyotes in New Mexico and another 1,424 in Arizona. There was considerable controversy about the vicious steel traps that would hold an animal by the leg until it starved to death. In the 1920s the survey began poisoning them with carcasses injected with strychnine, producing a slow and agonizing but certain death. But in Birdseye's time the big predators—wolves, coyotes, bobcats, and mountain lions—were still hunted or trapped.

According to Birdseye, the Biological Survey originally hired him as an "assistant naturalist." He said that he was part of a "scientific party, which was studying birds and mammals in New Mexico and Arizona." The U.S. Biological Survey sent naturalists throughout the territories by horseback to estimate animal populations and study their habits—especially what the large predators preyed on. They largely discredited the ranchers' claim that the wild predators were destroying livestock. While these animals did occasionally kill livestock, these domestic creatures were not their primary or preferred prey. In the case of coyotes, surveyors found that they mostly ate mice. But this did not stop the extermination program. In fact, one of the main missions of the 1973 Endangered Species Act was for the Department of the Interior to bring back all the animal populations it had destroyed at the beginning of the century with the U.S. Biological Survey.

Accustomed as he was to turning a profit from wildlife, Birdseye wondered what value there must be to the considerable numbers of killed coyotes. He asked at the Indian trading posts and was told that they were paying twenty-five cents apiece for the pelts from coyotes and bobcats. Surely, he reasoned, any fur must be worth more than that.

At the end of the summer he returned to New York and found a job with a New York insurance company and then with the city Department of Sanitation as a "snow checker"—the man who maintains records on the amount of snow removed from city streets after a blizzard.

But one of the first things he did upon his return to New York was to talk to furriers. He discovered that they would pay $1.25 each for bobcat and coyote pelts. He wrote to the traders to whom he had spoken in the Southwest and offered them fifty cents for pelts, allowing them to turn a 100 percent profit, and he then sold them in New York for $1.25. Even with shipping costs he earned $600 on his fur trading that winter. He returned to the Southwest for the U.S. Biological Survey in the summer of 1909.

In the Southwest he acquired some food habits that remained with him all his life. He developed a taste for the exotic. When Birdseye found something in nature, he always wondered what it would taste like and what would be the best way to cook it. This probably began with his natural curiosity, but he discovered that he enjoyed eating unlikely foods.

He liked his rattlesnake cut in slices, dusted with flour, and then fried in salt pork. While they were camped on one of the still-undeveloped rims of the Grand Canyon, the only provision the naturalists had for dinner was pork belly, which they discovered had gone rancid. Not to worry, Bob Birdseye would find dinner. He gathered field mice, chipmunks, gophers, and even a few pack rats. He carefully skinned and gutted the little bodies and wrapped them all in cheesecloth. Then he simmered it in a pot of boiling water. He praised the resulting stew, though there is no record of anyone else sharing his enthusiasm.

His colleagues rarely partook of the Birdseye specialties, preferring the canned food—mostly beans, corn, and tomatoes—that they took with them as they traveled the countryside by packhorse. Birdseye was struck by the extent to which westerners lived on canned food. Traveling by horse, he would see something sparkling in the sun on the horizon and realize they were approaching a ranch or a town, which was consistently marked by the piles of discarded cans. Though he always regarded canning as an inferior way of preserving food because it had to be heated and was no longer in its fresh state, and while he played a major role in the decline of canned food in the American diet, he personally had a great fondness for it. "He just loved canned food," said Gypsy, the wife of his eldest son, Kellogg. It was the food of his youthful adventures and filled him with nostalgia.

In the winter of 1909–10, Birdseye strengthened his ties to the U.S. Department of Agriculture with a bureaucratic job in the capital. At the time the secretary of agriculture, James Wilson, a McKinley appointee, had held the post for thirteen years. William Howard Taft, Wilson's third president, had retained him, and he stayed until the Woodrow Wilson administration in 1913. He remains the longest-serving cabinet officer in American history. Under him the Department of Agriculture was greatly expanding its gathering of data on wildlife, farming, horticulture, and forest protection, and there were many opportunities for an ambitious and adventurous young man such as Clarence Birdseye. But although Birdseye was looking for opportunities, he appeared to be more interested in those that involved adventure rather than career building.

In 1910, Washington, D.C., was a genteel southern town, with horse-drawn carriages and hitching posts still common. But construction was booming, and it might have become a high-rise megalopolis like Manhattan, except that year a law was passed restricting building heights to the width of the adjacent street plus twenty feet.

It is not surprising that a smart and adventurous young man from a distinguished New York family would meet the geographer Samuel S. Gannett. Like C. Hart Merriam, the head of the U.S. Biological Survey, Gannett was one of the thirty-three founding members of the National Geographic Society.

The society had been formed in 1888, led by Gardiner Hubbard, a Boston lawyer who had financed Alexander Graham Bell's development of the telephone and had been the first president of the Bell Telephone Company. When the society was founded, Hubbard had declared, "When we embark on the great ocean of discovery, the horizon of the unknown advances with us wherever we go. The more we know, the greater we find our ignorance. Because we know so little, we have formed this *Society* for the increase and diffusion of geographic knowledge."

This was Birdseye's kind of language. By 1910 the society's magazine had distinguished itself for exploration, investigating new ideas, and photography—all passions of the young Birdseye. Under its editor, Gilbert H. Grosvenor, a former Amherst history professor, the magazine avoided all political controversy. This too was Birdseye's style. The rest of the Birdseyes were not apolitical. His daughter Ruth would be active in the League of Women Voters, for instance. His own children and their spouses never knew if Birdseye was a Democrat or a Republican, even though they were mostly loyal Democrats. On the other hand, his father had been an outspoken Republican. But there is no record of Birdseye ever uttering a political statement. He was not a rebel. He never denounced religion, war, or corporations, but his participation in such institutions was always slightly removed. At heart he was something of an outsider, the odd kid they called Bugs, a likable maverick who knew how to get along with people. And so he had something in common with the National Geographic Society, adventurers who tried to remain apolitical at the center of the Washington establishment.

The *National Geographic Magazine* at the time Birdseye came to Washington was growing in circulation because photography was becoming increasingly popular, and it was known for its excellence in a certain type of documentary and anthropological photography. In 1907 it had created a sensation with Edward Curtis's portraits of American Indians. Birdseye, like many of his generation, was enthralled by the possibilities of photography. In fact, he harbored dreams of becoming a photographer, going to far-off exotic places, and recording the landscape and customs.

Samuel Gannett was the younger brother of another of the original thirty-three National Geographic Society founders, Henry Gannett, chief geographer for the U.S. Geological Survey, an organization whose topographical maps of the American West in the late nineteenth century set a new standard for mapmaking. Gannett used longitude and elevation from sea level for charting these maps, which made them not only handsome maps but also more accurate than anything preceding them. The Gannetts, like the Birdseyes, were of old New England stock. In fact, the first Gannett arrived in 1638, one year after the first Birdseye. Henry and Samuel were raised in the boatbuilding port of Bath, Maine. As a topographer, Henry Gannett traveled in some of the roughest terrain in the American West and was one of the first to ascend to the top of Mount Whitney in California. In 1906, shortly before Birdseye went west, the highest peak in Wyoming, Gannett Peak, in the northern Wind River Range on the Continental Divide, was named for Henry Gannett, one of the first to ever climb it. Today it is still considered the most difficult peak to climb in the continental United States.

Henry's younger brother Samuel was also a geographer with the U.S. Geological Survey, and he accomplished numerous feats of exploration. The West still had uncharted territory, and as a surveyor Samuel Gannett defined the borders of several new western states.

Among the wildernesses opening up to exploration were the frozen lands of the North. When Birdseye arrived in Washington in 1909, Admiral Robert E. Peary had announced that he had been the first man to reach the North Pole. There was considerable controversy because a physician who had served on earlier Peary missions, Dr. Frederick Cook, claimed to have reached the North Pole on his own expedition the year before. It seems certain that Birdseye was following the exploration and the controversy because it fell on Henry Gannett to verify the claim and of course it was all covered in the *National Geographic Magazine*. Peary's expedition had shot color photographs with a process invented in 1907; the magazine published the photos.

Peary was already a famous arctic explorer who had led expeditions in Greenland and other far northern territories and was known for learning Inuit ways of dress and survival. He traveled by dogsled and was famously photographed wearing native-style arctic furs. Another remarkable thing he was known for was that his wife sometimes accompanied him on these arduous journeys.

All of this must have seemed exciting to young Birdseye, and much of the Peary story was to later be mirrored in Birdseye's own adventures. But what particularly interested him at the time was Samuel Gannett's daughter, Eleanor. Little is known about their courtship. Politics was not the only thing Birdseye didn't like to discuss. He did not talk about girls or his love life, nor did Eleanor talk much about such things. She was the shy one of the couple.

When they met, she was finishing her B.A. degree at George Washington University, an unusually well-educated woman for her generation. They dated for five years, during most of which they didn't see each other, because he was constantly off on his adventures. Then they married, and for forty-one years had the most unshakably devoted relationship. Could

Eleanor, coming from the world of explorers and inventors, see what he would become?

He didn't talk about being an inventor; in fact, even after he became one, that was seldom a word he used to describe himself. At least she could have seen that he had an insatiable curiosity and a great sense of adventure. He always wanted to understand how things worked and what would make them work better, and he was always looking for how to turn things others thought were waste into a profit. She thought he was brilliant, an impression that grew over the years. Life with him would probably be like how it was with her father and her uncle. And Birdseye could see that despite her quiet ways, Eleanor did not come from the kind of conventional home that he did, and she was not looking for the kind of Birdseye who would become a prominent lawyer or a Wall Street businessman. She was a woman who would accept and understand her husband's taking off on expeditions.

This was apparent because she was not put off by his habit, even when they were first dating, of leaving for distant destinations, of running off on adventures. By the spring of 1910, the Department of Agriculture had something interesting for him to do.

Fig.1.

32
40
36
38
24
10
19
17
12
14
15
15

Chapter 4

Ticks

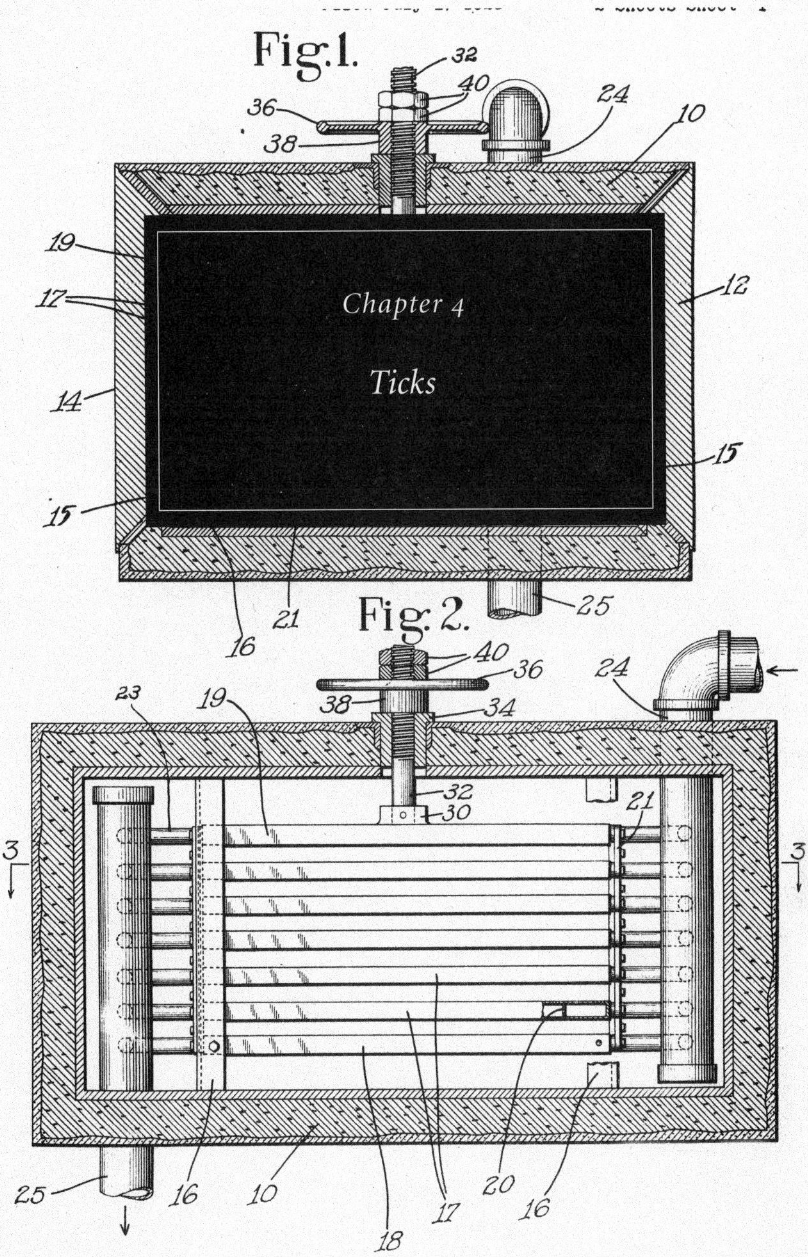

16
21
25

Fig.2.

40
36
38
34
24
23
19
32
30
21
3
3
25
16
10
17
20
16
18

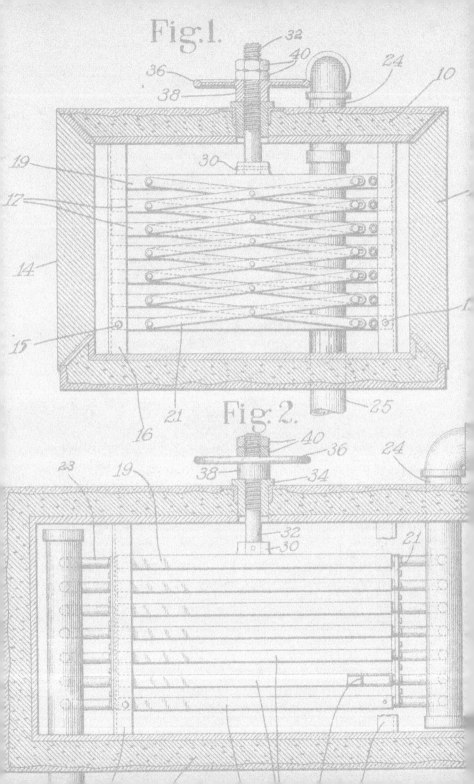

Fig.1.

Fig.2.

Nothing shows more clearly the distance between the real Clarence Birdseye and the image he liked to present to the public than his experiences in Montana in the spring and summer of 1910. According to Birdseye, he found this great job as a hunter for a medical project on Rocky Mountain spotted fever. To collect ticks, he was allowed unrestricted hunting of wildlife—permitted to shoot far over the normal limits. His was always a tale of adventure with a few good yarns thrown in. What he almost never talked about was that the campaign against spotted fever was extremely important and extremely dangerous; that he took a job few would have been willing to do; and that he played an important role in research that is a significant chapter in medical history, has saved many lives, and advanced understanding of a whole family of deadly diseases.

Rocky Mountain spotted fever, *Rickettsia rickettsii*, is a life-threatening disease. Much is still unknown about how it spreads and how it acts on the human body. But in 1910 nothing at all was understood about it. It was a uniquely American disease, occurring only in the Western Hemisphere. Rocky Mountain spotted fever was its local name. In other places it had other labels, such as São Paulo fever, New World spotted fever, and American spotted fever. But it was first identified in the Rocky Mountains and so labeled. It has also been found in

Brazil, Colombia, Costa Rica, Mexico, and Panama. There are tremendous differences in the mortality rate from the disease in different areas, and the reason for this remains a mystery. In one area it may kill 5 percent of victims and in another 70 percent. Typically, an outbreak of Rocky Mountain spotted fever will kill one in five victims, a high enough mortality rate to terrify a stricken population.

A victim can have the disease for as long as two weeks without exhibiting any symptoms. Then there is a sickish feeling for a few days resembling the signs of an oncoming cold or flu. Then the patient often experiences an extremely severe headache and painful joints. Light begins to hurt the eyes, and the neck becomes stiff, accompanied by dangerously high fever. At this point doctors sometimes suspect meningitis. The patient cannot sleep and at times becomes delirious. After several days of these symptoms, slightly raised red spots appear on the skin, easily confused with measles. We now know that the organisms known as rickettsias invade the cells lining the capillaries, causing them to swell and break, and blood seeps through the capillary walls. These tiny welts show up on the entire body, including the palms of hands and soles of feet. Without any treatment whatsoever, many people will recover in a matter of weeks. But a significant number, especially over the age of forty, will die.

There was probably a nineteenth-century history of the disease. But its pathology was not understood. There are records of an outbreak of something called black measles during the Civil War, and there are accounts of diseases with similar names among westward-moving nineteenth-century pioneers. But until the U.S. government began researching the illness in the beautiful and wild Bitterroot valley of western Montana in 1910, there had been no formal scientific study of the disease. At that time scientists had only recently identified Rocky Mountain spotted fever as a distinct illness. It had often been confused with European rash-producing diseases such as typhus, the plague of the Napoleonic Wars.

Birdseye had been born into not only a new age of invention and technology but also a new age of medicine. It began with Louis Pasteur, the French chemist who in the 1850s demonstrated that fermentation was caused by the growth of microorganisms, living things too small to be seen by the naked eye. This led to the process of protecting milk from spoilage that bears his name, though his U.S. patent was for applying the technique to beers and ales. But it also led to something much bigger. Pasteur proposed that microorganisms caused diseases. As with all great discoveries he did not arrive at it in a vacuum. In 1546, the Italian physician Girolamo Fracastoro claimed that seedlike creatures caused epidemic diseases. In 1835, another Italian, Agostino Bassi, demonstrated that a powdery substance that, in fact, was thousands of living organisms was killing silkworms. In 1854, John Snow demonstrated that an outbreak of cholera in the Soho section of London was being caused by something in the drinking water. A new field was emerging, first developed by the German doctor Rudolf Virchow in the mid-nineteenth century and known as cellular pathology—the study of the role of microscopic cells in the development of diseases.

But Pasteur was a prominent figure and explained his pathogenic theory of medicine well, and it launched an international debate about "Pasteur's germ theory." American doctors did not think much of the germ theory. Weren't the French always having theories? But in 1883, three years before Birdseye was born, the entire approach to medical science began to change because a German scientist, Robert Koch, demonstrated that a microscopic creature, a microorganism that he named *Vibrio cholerae*, caused cholera. The germ theory was right. The following year this organism was found in the harbor water in Marseilles during a cholera epidemic. Armed with this new knowledge, medical researchers began traveling the globe conquering epidemics. Cholera was often eradicated, including in New York City, by purifying the drinking water. Viruses were identified in outbreaks of numerous diseases, including bubonic plague, typhoid, leprosy, and tuberculosis. During the 1898 Spanish-American War, Walter Reed, a U.S. Army physician, went to Cuba and found that a microorganism carried by a specific species of mosquito caused yellow fever, which was decimating American troops. Yellow fever became manageable through mosquito control, making the Panama Canal project viable.

These were romantic figures, these scientists and doctors who traveled to remote places and conquered dread diseases. It was dangerous work, and some died, including a close coworker of Reed's in Cuba. In 1910 the U.S. government decided to send such medical crusaders to the Bitterroot Mountains of Montana, and among them was Clarence Birdseye.

The Bitterroot valley is a hundred-mile stretch of wooded flatland, once a lake bed, between the Bitterroot Mountains and the Sapphire range. The Bitterroot River runs through it, an agreeable spot sheltered by the mountain ranges. The Salish, or Flathead, Indians, who gave the place its name from the bitter roots they ate from a pink wildflower, were a healthy people. The Lewis and Clark expedition, which made a point of reporting on local medical conditions, had no diseases to report from its 1805 visit to the Bitterroot. The whites started settling there in the mid-nineteenth century with no particular health problems. The first documented case of Rocky Mountain spotted fever in the Bitterroot valley was in 1873. The patient was one of the few white men living there, and he died of the disease. In the late nineteenth century the widespread occurrence of this strange disease coincided with the spectacular expansion of the lumber industry. Was the disease related to the chopping down of trees? The west bank of the river, where the lumbering was done, was also where the disease was most prevalent.

The significant advances in medicine and epidemiology led governments to conclude that they could master the spread of disease, and so in the second half of the nineteenth century there was a huge upsurge in public health measures, which reached the Bitterroot valley at the turn of the century. In 1901, Montana established a state board of health. But there had already been a Ravalli County board of health since 1896.

Once agencies were established to monitor disease, they started noticing a large number of patients in the Bitterroot valley with diagnoses such as black typhus fever, blue disease, and black fever. These diseases were not unique to the Bitterroot valley. There were also outbreaks in neighboring Idaho, in Colorado, Oregon, and Wyoming, and possibly earlier in California, Washington, Nevada, and Utah. But it was a much more severe illness in the Bitterroot than anywhere else. Colorado was thought to have a mortality rate of 23 percent, yet Montana had fourteen cases in the spring of 1901 and all but four died.

People in the Bitterroot were becoming terrified of this deadly disease. This was the age when the West, western photography, and Indian art were becoming fashionable, and westerners were beginning to recognize the economic potential of tourism. The Bitterroot valley, rolling flatlands of native pine and fruit orchards below the jagged crests of the high Rockies, was one of the most beautiful spots in the region. It had been a fruit-producing area since 1871, when the first apple seeds were mailed from Plymouth, Massachusetts. Fruit and dairy agriculture was prospering, and land values were rising. Special trains from Chicago carried prospective eastern investors to look over the valley. The Bitterroot, it was believed in Montana, was poised for a boom. But tales of a deadly disease were crushing such plans.

The leading theory was that these black measles were somehow caused by melted snow—a fusing of the fact that the disease began with the thaw and the fairly new discovery that many plagues were waterborne.

In 1889, a particularly well-educated Montana physician who had studied the work on yellow fever noticed a tick on a patient and suggested that this might be significant. In 1902 a new breed of investigators turned up. Since the disease never struck before mid-March, they arrived in the spring. Two pathologists from the University of Minnesota, Louis B. Wilson and William M. Chowning, headed the new team. Wilson and Chowning shocked both the public and the medical community by documenting eighty-eight cases that had occurred between 1895 and 1902, of which sixty-four were fatal—an unheard-of mortality rate of 72.7 percent.

The pathologists explored all the newly discovered sources

of disease, including water, insects, and arthropods—ticks and mites. In every case that Wilson and Chowning examined, the patient had been bitten by wood ticks two to eight days before symptoms appeared.

In 1904, with property values in the Bitterroot plummeting, the Montana State Board of Health obtained a commitment from the U.S. surgeon general to sponsor research in the Bitterroot until the nature and causes of Rocky Mountain spotted fever were completely understood. But one of the effects of this commitment was to emphasize that the answers were not yet known, and so with plans for tourism, irrigation, and orchards in the works the local press started to avoid mentioning the disease. There was a shame attached to it, and as in the AIDS epidemic seventy years later, when someone died of Rocky Mountain spotted fever, the obituary avoided giving the cause of death.

The locals, especially those who worked outdoors, adopted the habit of examining their bodies nightly for ticks They started carrying small bottles of carbolic acid with them while working outdoors, especially on the contagion-prone west bank of the river. Carbolic acid was one of the miracles of the new age. The British surgeon Joseph Lister, who had pioneered the idea of keeping surgery safe from dangerous microorganisms, developed it as an antiseptic. The problem was that if you applied it directly to the skin—the Bitterroot people would open the bottle and hold it upside down over the tick bite—it left a red acid welt. Still, this was better than the killer fever. Not surprisingly, given the level of fear, many fraudulent medicines guaranteed to save you, including sarsaparilla, were sold in the Bitterroot. This trade was much slowed down in 1906 by the passage of the Pure Food and Drug Act, a piece of legislation that Birdseye would one day play an important role in revising.

Among the new fields of science was entomology, the study of insects. After 1890, when it was discovered that arthropods, an eight-legged cousin of insects, could transmit disease-causing microorganisms, a wide range of doctors, public health officials, and veterinarians became interested in studying ticks. In 1909 the term "medical entomology" came into use for the specific study of the role of arthropods in disease. The Bitter-root project became cutting-edge medicine.

Howard Taylor Ricketts, a pathologist at the University of Chicago, became interested in Rocky Mountain spotted fever, and he successfully identified the organism that caused the disease, for which he won the dubious honor of having it named after him, *Rickettsia*. He then went off to study an outbreak of typhus, a somewhat related disease, in Mexico and died while working on it. Field epidemiology was very dangerous work.

As more and more disciplines and more and more government agencies got involved in the Bitterroot project, infighting developed. It was centered on the belief of the physicians that the new scientists, the pathologists and entomologists, were not really qualified and should defer to their greater learning and training. Physicians would sneeringly refer to entomologists as "bugologists." Many of the locals felt the same way and openly laughed at the hypothesis that the disease was caused by ticks. After all, carbolic acid hadn't worked.

In February 1910, Robert Cooley, Montana's official state

entomologist and a man deeply committed to solving the Bitterroot spotted fever problem, went to Washington to ask for help from the U.S. Biological Survey in studying the connection between wild animals and the life cycle of the tick. The Biological Survey and the Bureau of Entomology were both housed in the same small brick building in Washington, and Cooley left with three recruits for the Bitterroot. Willard V. King, a gifted entomology student from Montana, had delayed his senior year to work for the Bureau of Entomology. The U.S. Biological Survey sent Arthur H. Howell, a thirty-seven-year-old zoologist who would publish 118 works on birds and mammals by the time he died in 1940, and Clarence Birdseye, a twenty-three-year-old college dropout.

No one else had been willing to volunteer because the work seemed too dangerous. Howell, a married man with children, agreed to work in the field camp but only on condition that a younger volunteer, single with no children, would come with him to go trapping and shooting in the highly infested areas. This was Birdseye's kind of adventure, and he eagerly volunteered. According to legend, when C. Hart Merriam asked for volunteers in the Biological Survey, the reply was, "Why not ask Birdseye?"

An abandoned log cabin that had once been a farmhouse became the home and research center for Cooley, King, Birdseye, and a cook named Paul Stanton. They named it Camp Venustus, after the scientific name for the dangerous tick, *Dermacentor venustus*. The four men posed for a picture in front of the cabin, Stanton in his cook's apron and the other three in suits and ties. Birdseye in his tweedy three-piece suit, thinning hair, and spectacles did not look that different from the famous inventor and entrepreneur he would become in his fifties. The photograph gives no clue that this small man in the suit risked his life every day on horseback wandering into the canyons with his hunting rifle.

After World War II, Esther Gaskins Price, a former researcher at the Mayo Clinic, decided to write a history of the campaign against spotted fever in the Bitterroot. She lived in Marblehead, Massachusetts, only a short drive from Gloucester, where Birdseye was living, and so she went to ask him about the Bitterroot campaign. Typical of Birdseye, he said little about his significant contributions to the medical research. He was asked many times about this period, and he always talked about how many animals he killed, sometimes pointing out that it was well over the usual limit, and then he liked to throw in a few good yarns like the one about the calendar. The cabin had a calendar with a picture of a naked woman in a bathtub. Birdseye, who loved photography, documented all his work in the Bitterroot with photos that he submitted to Washington with his reports. King managed to shoot a picture that showed the girl without the calendar, so it appeared to be an actual voluptuous naked woman in a bathtub. In the foreground was a sign reading, "C. Birdseye's Office." This he slipped between photographs of mountain goats and bears that Birdseye shipped off to Washington. They spent a lot of time out in the wilderness playing practical jokes on one another, and King thought this would be a great joke on Birdseye. But he had misjudged Washington, and the photograph created such an uproar that King was nearly removed from the project.

The closest Birdseye ever came to talking about how dangerous his work had been was the story he told Price about the trick he and King played on Howell.

While the different agencies and the bugologists and physicians were all competing on this small turf, Birdseye and King could see that this was a young man's game. They would take all the risks while Howell would remain in the cabin writing a paper on their work on which only his name would appear. Howell had made clear from the outset that he intended to avoid danger. Birdseye and King remembered how back in Washington, Howell was the one who talked about how he was married with children and the job was too dangerous. While King and Birdseye had made arrangements with a nearby doctor in the event a tick infected one of them, Howell often stated that at the first sign of a tick being embedded on him, he was packing for Washington.

In the cabin the three checked one another for ticks every morning and night. So Birdseye, as he later told Price, proposed to King that "if on one of the nightly inspections, we could just discover a tick between Mr. Howell's shoulder blades where he could not possibly see that it was purely imaginary, he would be off for Washington and we could do the paper work as well as the field work." The idea, Bob told Price, had come to him while skinning a coyote he had shot and putting the ticks they had found into vials. They put one tick engorged on coyote blood in a separate vial. At night while examining Howell, Birdseye claimed to find a tick between his shoulder blades. Howell was so upset he insisted that Birdseye not only cut it out with a scalpel but also cauterize the wound, which Birdseye obligingly accomplished with a heated blade while Howell winced in pain. Then King handed him the vial with the tick from the coyote. It worked even better than the two young men had anticipated. First thing the next morning Howell was making arrangements to return to Washington.

Some weeks later Birdseye had a remarkable hunting trip during three days of pouring rain. He shot three wild mountain sheep and a brown bear and collected an enormous quantity of ticks from the bodies. He was so excited that he neglected to check himself for ticks, and when he finally did by an evening campfire, he found seven—three of them thoroughly engorged with his blood. For several days he had the aches and nausea that were the well-known symptoms of spotted fever. He spent three terrified days before he realized that he simply had a grippe from being out for days in the rain. But it left him feeling that his joke on Howell was not as funny as he had thought it was.

Bob Birdseye liked and got along with most people, even Howell. Years later he told Howell what he had done, and according to Birdseye, Howell laughed. He had probably been happy to return to Washington anyway.

Now the two young men, Birdseye and King, were on their own, along with the cook Stanton and occasional visits from Cooley. Their task was to gather ticks to be studied, and an unknown percentage of these tiny creatures could deliver a fatal bite. There was the possibility of taking an untested horse serum to possibly protect them from the disease, but Birdseye and King thought this as risky as facing the ticks.

While they were in the field, Josiah J. Moore, a University of Chicago pathologist, concluded from experiments that the minimum amount of time that a tick needed to attach to a guinea pig in order to infect it was one hour and forty-five minutes, and on average it had to stay in place, feeding on the host, for ten hours. Based on this research, Cooley instructed Birdseye and King to stop their work every two hours to inspect each other's bodies for ticks. They did occasionally find a tick, but neither ever contracted spotted fever, either because of luck or because Moore was right about the necessary feeding time.

Because of Birdseye and King this became a standard procedure in tick collecting for spotted fever. The clothing they devised also became standard. They wore high shoes to which cloth leggings were attached and fastened to their calves by drawstrings. Their outer clothes, cotton for the summer weather, were soaked in kerosene, which at least for a short time served as a tick repellent. At night they left their clothes in an airtight closet with carbon bisulfate, a common insecticide.

King worked late into the night at his microscope. By day he went into the brush with his collecting flag and vials, while Birdseye, with a hunting rifle on his shoulder and as many traps as he could carry, went out into the most infested canyons. King had a white wool flag on a pole, later replaced by lighter flannel, and he waved the flag in the brush to attract ticks. Birdseye photographed the animals, and King photographed and studied the ticks. At night the two young men lay in their cabin having lively debates about the future of the spotted fever campaign. They wanted to document the complete life cycle of the tick and determine what animals served as hosts at which stages.

Birdseye shot and trapped gophers, chipmunks, pine squirrels, woodchucks, ground squirrels, wood rats, snowshoe rabbit, cottontail rabbit, several species of mice, flying squirrels, badgers, weasels, muskrat, and bats. He also shot and killed the large and dangerous brown bear, the elusive mountain sheep found in remote and rugged rock ledges, coyotes, both mule and white-tailed deer, and elks. He shot most of what he saw, but he also set traps and collected 4,495 ticks. In the process he killed 717 wild animals, though he had said his goal was to get 1,000. He was Buffalo Bill at last. (According to Price, he killed over 1,000 animals, but the 717 figure from the project seems more likely. Apparently, Birdseye exaggerated the kill when talking years later to her.) He had been exempted from the normal hunting restrictions. Still, when the game warden J. L. DeHart learned of the quantity of animals he had killed, he angrily referred to it as "wanton slaughter."

But the slaughter was not wanton. The insects that were collected led King to discover that the tick had a two-year life cycle and not a one-year cycle, as Ricketts had thought. This was significant because it enabled Birdseye to determine that the tick fed on small animals such as mice when it was young but as a mature adult moved to large animals. Birdseye showed that it was pointless to have a campaign only against small rodents, including the gopher, correctly called Columbian ground squirrel, which had come to be regarded by the pathologists as the leading culprit. Birdseye and King concluded that spotted fever could be controlled if a campaign against small rodents were combined with a program treating the large animals, principally domestic livestock, with repellents. It was an important breakthrough. This was not what ranchers wanted to hear, because they did not want the expense of treating their livestock and had liked having everything blamed on the little gopher that was for them a hated destructive pest.

Now Cooley erected livestock-dipping vats throughout the

valley for ranchers "voluntarily" to run their cattle through. But the program was not really voluntary, because ranchers who did not dip their livestock faced quarantine on their animals. Adding to the unpopularity of the initiative, in a community that tended to be distrustful of any government programs, it took some experimenting to get the right strength of arsenic in the dipping vats so it would kill the ticks but not hurt the livestock. Hides and udders got burned before the right formula was found. The ranchers had been much happier with Ricketts, who simply endorsed a program to exterminate gophers.

In the fall the ticks and the disease vanished, and the study packed up until the following spring. Birdseye probably stayed with the Biological Survey and went back to Washington, where he could spend time with Eleanor. According to a journal in Birdseye's handwriting, he remained in the Bitterroot for at least part of the winter of 1911 trapping and experimenting with various poisons, even though there were no ticks and therefore no medical project during those months. Perhaps he was taking advantage of the fact that winter is an excellent time to lay poisoned bait because food is scarce. In times of plentiful prey most wild mammals will not eat carcasses. In his journal he repeatedly explained that he was exterminating animals at the request of local farmers. In the spring of 1911 he returned to his dangerous work with wood ticks.

Birdseye's ideas for a program to control spotted fever were laid out in a report initially published by the Biological Survey in 1911 with Henry Wetherbee Henshaw, the head of the survey, as the first author and Birdseye as the second. A more extensive report came out the following year in the

Department of Agriculture's *Farmers' Bulletin* with Birdseye as the sole author. The report borrows a little from the newly emerging medical knowledge of spotted fever and has some of what King learned of ticks, but most of it is pure Birdseye. The report, *Some Common Mammals of Western Montana in Relation to Agriculture and Spotted Fever*, reveals a great deal about both Birdseye and the U.S. Biological Survey. The report contains no notion of the need to preserve nature or the natural order. The larger issue was how nature could best serve human endeavor, in this case agriculture. This was the point of view of the U.S. Biological Survey, which, after all, was a subdivision of the U.S. Department of Agriculture. The report called for the extermination—that was the word used—of gophers, chipmunks, ground squirrels, pine squirrels, woodchucks, rabbit, and badgers in western Montana. Birdseye discussed the life cycle and habits of these animals with the thorough observation and insight of a trained and experienced biologist. He then explained how best to kill them—when to shoot them, when to trap them, when to poison them, which poisons to use, what time of year to do so, and how to apply the poisons for each species. While he did discuss each species's relationship to ticks and spotted fever, a great deal more attention was paid to the destruction to agriculture caused by these unwanted animals. Although he argued that his and King's research showed the Columbian ground squirrel, a.k.a. the gopher, to be the most important host to young ticks, he goes on to point out that the gophers' habit of eating growing grain and garden vegetables and burrowing in planted fields "is sufficient to warrant their destruction." Pocket gophers, on the other hand, are seldom tick hosts, and Birdseye stated that their relationship to spotted fever couldn't be of much importance. But then he goes on to point out that because they destroy irrigation ditches, kill fruit trees, damage hay fields, and eat garden vegetables, they ought to be exterminated too. Weasels were to be spared because they killed so many rodents. To prove his point—always a scientist testing a hypothesis—he kept a weasel in a cage and periodically placed the largest gopher he could find in with it and observed the weasel devour the newcomer. He felt this way about badgers as well, that they were great rodent hunters, but reluctantly conceded that they did host a lot of ticks so should probably be killed.

Birdseye wanted to spare big game. He argued that though he found mountain sheep, elks, and brown, black, and grizzly bears infested with ticks, they did not have enough contact with humans to be considered significant carriers. One notable exception, true to the policies of the Biological Survey, was the coyote. "It is important that their numbers be reduced."

He illustrated his points with his photography. The cover was graced with a shot of an apple tree killed by pocket gophers. Then there were the unappealing photos of "twenty-seven ground squirrels, chipmunks, and mice killed by 7 cents worth of poisoned grain." There are carrots ruined by gophers, the washout of a ditch caused by gophers, and a photograph of his horse carrying two hundred animal traps.

Nature was to be put in the service of man, and nothing in it should be allowed to impair food production. Nowhere in this study was there a sense of preserving the natural order or the concept that removing so many species would narrow biodiversity and impair the entire ecosystem. If Birdseye was a self-taught biologist, he was not an ecology biologist—a biologist who studies ecosystems. The discipline existed in his day, but it was not as fashionable as it is today. Today the

U.S. Biological Survey has become the Fish and Wildlife Service, which, though not without its controversies, is expected to protect and preserve nature.

In the winter of 1912 Birdseye was back in Washington. He did not return again to Montana. The spotted fever campaign continued without him and had its first fatality, Thomas McClintic of the U.S. Public Health Service. In all, fourteen people died in the fight against spotted fever between 1912 and 1977. With the discovery of antibiotics in the 1940s, a cure was found for spotted fever and the crisis seemed to be over, but then there was another serious outbreak in the 1970s. The early research that Birdseye participated in has led to solving riddles about other serious illnesses such as Legionnaires' disease, Lyme disease, and Potomac horse fever. It led to the second-largest government laboratory in the United States being established in Hamilton, Montana, where, during World War II, studies on insect-borne diseases were to save the lives of soldiers throughout the Pacific.

If Birdseye had done nothing else, his fieldwork on spotted fever and ticks would have earned him a footnote in history as it did Willard King, who went on to distinguish himself in studying the mosquitoes in New Guinea that were plaguing World War II troops.

But Birdseye had a different destiny, though he had little idea of his future at the time. He only knew that the century was still young and so was he, and there were opportunities for other adventures. All his life Birdseye was restless—eager to move on to the next thing. In 1912 the next thing was Labrador, a sparsely populated frozen wilderness.

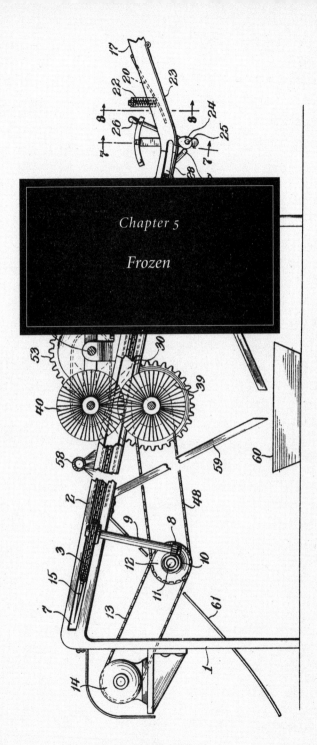

Fig. 3.

Chapter 5

Frozen

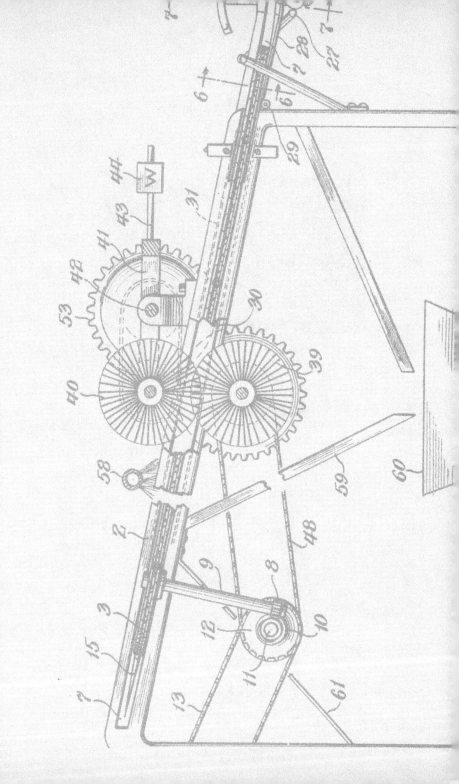

The coincidences of destiny are unpredictable things. Had Clarence Birdseye not been invited to spend six weeks along the Labrador coast on the hospital ship of the celebrated medical missionary Wilfred Grenfell in 1912, a brief and seemingly inconsequential chapter in Birdseye's life, he might have remained a curious minor figure, and this biography would not have been written. Or maybe he would have done something else no one had ever thought of.

He was looking for adventure but also opportunity, and he hoped that perhaps Labrador would present an opportunity of some kind. He said good-bye again to Eleanor, the explorer's daughter who understood his need to go, and he headed north.

Labrador at the time was a possession of Newfoundland, itself a dominion, with the same self-ruling status in the British Commonwealth as Canada. Like much of maritime Canada and New England—for that matter, like Gloucester, Massachusetts, which would later be Birdseye's home—Labrador and Newfoundland's first Europeans were fishermen in the seventeenth century, drawn by the huge profits to be made in Europe from the dried and salted codfish with which the American North Atlantic was teeming. Cod was another example, like the buffalo and the passenger pigeon, of the extraordinarily large populations of North American wildlife beyond anything Europeans had known.

But Europeans had also never known winter like Labrador had. Both its duration and its coldness were beyond their experience. Many Europeans thought the winter, even in New England, was so harsh as to make the region uninhabitable for year-round settlement—an odd assessment since there were indigenous people living there. But New England was far milder than Newfoundland and Labrador and soon had year-round settlements and even a year-round cod fishery. In Newfoundland and Labrador, on the other hand, the harbors freeze up in the winter, and fishing had to be discontinued. A few tough fishermen would survive the winter months there, and the rest would leave and return for the spring thaw.

If Newfoundland was a backwater frontier, Labrador was its wildest uncharted wilderness. In the early eighteenth century, when France still had a large presence in the American North Atlantic, a French map showed the Gulf of St. Lawrence and Newfoundland. It also charted with some accuracy portions of the Grand Banks, the shoals at sea where codfish were caught. At the top of the map was a barely shaped lump of land where Labrador is labeled "Pays des Esquimaux," Eskimo country. Neither the French nor the British did much to establish themselves there.

Some fishermen did settle there, but Newfoundland and Labrador never developed the economy or the population of prosperous New England. This is why there was little interest in separating from England in 1776; in 1849, when Canada gained self-government; or even during the second half of the nineteenth century, when the Dominion of Canada gained territories and grew into a sizable country. Even when Newfoundland and Labrador did contemplate self-rule, it was as its own country and not part of Canada. To the people of

Labrador, "Canady" seemed a very far-off and different place from these cold and rugged colonies. In 1907, Newfoundland gained dominion status, and Labrador went with it, making Labrador a kind of colony of Newfoundland without representation in its legislature. The two didn't join Canada until 1949.

When Wilfred Grenfell was studying Labrador, he found an old map that said, "Labrador was discovered by the English. There is nothing in it of any value." A visitor might easily have that impression, but for centuries Labrador and Newfoundland had been viewed by Europeans, who never went there but saw their countrymen earning fortunes on cod and furs, as sources of great wealth. But the people in Labrador who produced the salt cod and fox furs for Europe were pitifully poor, among the poorest people in all of the Americas.

Labrador had meat that could be shot and hides to be cured and salmon, halibut, herring, lobster, and, above all, cod to be caught and eaten in the summer. In March there were seals to be hunted. But anything else that was to be consumed, including fruits and vegetables and grains, had to be imported. Berries were the only local fruit, and the only Labrador-grown greens were the tops of turnips in the summer. The fishermen and fur trappers traded their products to large companies like the Hudson's Bay Company in exchange for supplies and ended up in perpetual debt. Grenfell told the story of a man who, having failed in the local salmon run, wanted to row to cod grounds but had to trade his anchor for provisions in order to make the trip.

In 1912 there were about 250,000 people living in Labrador and Newfoundland. Most of them were fishermen in the more-southern island of Newfoundland. Population in Labrador was sparse with no large towns or cities. Most Labradoreans were variously fishermen huddled in the rocky coves along the coast; fur breeders, trappers, and traders; the indigenous Inuit, known to white people as Eskimo; or small groups of Montagnais Indians, who lived in the interior, descendants of the Algonquin, who had been pushed north.

There was little electricity and few roads, dogsleds were the only winter transportation except for hiking in snowshoes, and there was no telephone service. Most people made their own soap from boiling spruce ashes with seal fat. There were no doctors outside of two small hospitals too far for most people to get to, both built by Grenfell.

The few opportunities for medical attention in Labrador were provided by Wilfred Thomason Grenfell. Grenfell was born in 1865 in Parkgate, a coastal village in Cheshire by Liverpool Bay, across the Dee from the green hills of North Wales. His father was the headmaster of a boys' school. Fishermen seined for salmon there, and Parkgate was famous for its mussels and its shrimp. Wilfred grew up talking to fishermen who liked his friendly manner and the way he so admired them. He was a sports fanatic who also loved sailing and the sea. He and his brother refitted an old fishing smack and would take to sea without navigation skills, or, as he put it in his autobiography, "without even the convention of caring where we were bound so long as the winds bore us cheerily along."

A social conscience was valued in the Grenfell family, and they all learned about Wilfred's great-grandfather who worked with William Wilberforce, leader of the British movement to abolish slavery.

In 1883, Wilfred began medical school in London. In 1885, the second-year medical student was returning from seeing a patient when, on a whim, he went into a tent where the famous American evangelist team of Dwight L. Moody and the strong-voiced gospel singer Ira Sankey, composer of numerous popular hymns, were appearing. Most everyone in Britain, especially young people, had heard of Moody and Sankey, and their revival meeting was something that everyone should see at least once, so Grenfell, who was ambivalent about religion, stepped in.

An unknown speaker was delivering a long, tedious prayer, and Grenfell was about to walk out when a heavyset, thick-bearded Moody said, "Let us sing a hymn while our brother finishes his prayer." Grenfell thought this was a remarkably dexterous and gentle way of dismissing the boring speaker, and he was impressed by Moody's finesse. He went to more meetings and was finally swept into the movement by a group of Moody followers known as the Cambridge Seven. These were star athletes, and there was no one whom Grenfell admired more than great athletes. Two of these young men played cricket for England. At one of their meetings Grenfell committed his life to Jesus Christ. Moody's message fit Grenfell's own cynical view of religion, that you could either be a hypocrite spouting religion or help people in keeping with the teachings of Jesus. Moody encouraged missionary work around the world, especially in China. That was why he was popular with the young men of England. What was a young man to do in Victorian England now that the empire was secured and there weren't even that many wars anymore? You could go out as a missionary. Many did, Grenfell included.

In his autobiography, Grenfell wrote that Moody helped him to see "that under all the shams and externals of religion was a vital call in the world for things that I could do." Years later he named one of his sled dogs Moody. Moody the dog would be drowned saving Grenfell's life when he was trapped on an ice floe.

After receiving his medical degree, Grenfell joined the Royal National Mission to Deep Sea Fishermen, a group that tended to the needs of commercial fishermen in the North Sea. The North Sea fishermen of the time were being exploited by copers, floating entertainment centers providing alcohol and tobacco at exorbitant prices. (Sometimes fishermen lacking cash paid those high prices in fish.) Grenfell's ship brought religion and medicine but also—it now seems odd for a doctor—provided fishermen with tobacco at cost to undersell the copers. Grenfell loved the North Sea fishermen for their camaraderie, their resourcefulness, their courage, and their lack of pretensions. And of course he was awed by their navigational skills. "At first," he wrote, "it was a great surprise to me how these men knew where they were, for we never saw anything but sky and sea."

But by 1890 the new steam-powered net trawlers were already causing overfishing in the North Sea, and the mission moved on to south Irish spring fishing and then to fishing off Scotland.

In 1892, Grenfell was sent to Newfoundland and Labrador to continue the work with fishermen. From the moment he first saw it, he was struck by this remote dominion. His crossing followed almost the exact same course as John Cabot's 1497 voyage of discovery on a ship that, as he pointed out, was not much bigger than Cabot's. He always remembered the excitement of first seeing the rocky cliff, the thousands of marine birds, whales cavorting in the sea, the whitecaps rolling into the coast, and huge shoals of fish everywhere, so thick that their wet black backs splashed on the surface of the water. He had never before seen so many fish, but what left the strongest impression on him was a man in a makeshift little boat who came alongside and asked if he was really a doctor.

The fisherman took Grenfell to a hovel made of sod, "compared with which the Irish cabins were palaces," and showed him a man dying of pneumonia. Grenfell could not take him away on his hospital ship, because he was the family breadwinner and they could not afford to have him away during the few months in which they earned their living fishing before the harbors iced over. And so the man died. Grenfell realized that he had found the place where he was most needed.

Most of the fishermen had never seen a doctor. He wrote:

> Deformities went untreated. The crippled and blind halted through life, victims of what "the blessed Lord saw best for them." The torture of an ingrowing toe-nail, which could be relieved in a few minutes, had incapacitated one poor father for years. Tuberculosis and rickets carried on their evil work unchecked. Preventable poverty was the efficient handmaid of these two latter diseases.

The following year he built Labrador's first hospital in Battle Harbour, a few dozen dark-roofed white buildings wedged between the frozen sea and the icebound cliffs on the Atlantic coast of southern Labrador, just north of the straits that led to the mouth of the St. Lawrence. Then he built a second—it was little more than cottages—farther up the coast at Indian Harbour. In 1894, he formed the Grenfell Medical Mission. In 1899, its new hospital ship, the coal-fired *Strathcona I,* was launched. It had two masts and could also burn wood if coal was not available. It was named for the man who gave the ship to the mission, Lord Strathcona, whose real name was Donald Alexander Smith. Smith's place in Canadian history was established in 1885, when he symbolically united Canada by driving the final spike into the Canadian transcontinental railroad in the Rocky Mountains.

For Grenfell, as for Birdseye a few years later, Labrador offered the romance of adventure—an unexplored land where Stone Age relics could be easily found along with remnants of ancient whaling stations built by mysterious Basques. Grenfell wrote, "It is so near and yet so far, so large a section of the British Empire and yet so little known, and so romantic for its wild grandeur, and many fastnesses still untrodden by the foot of man!" This is what drew Birdseye as well.

Grenfell spent his winters, when Labrador ports were frozen and inaccessible, raising money in Britain and the United States. But in time he learned to love the Labrador winter, in the same way Birdseye would learn to love it: for its stark beauty, the thrill and fun of dogsledding, the simplicity of life without stores or crowds, and the warmth of the rare encounters with the few comrades spending the winter the same way. Grenfell wrote in *Tales of the Labrador* that because of the absence of stores "we are relieved of the constant suggestion that we need something."

Starting in 1902 he spent every other winter in Labrador. Once spring came, he was back on his hospital ship bringing medical help, sailing the Labrador coast into its deep unknown fjords and islands, some of which had never been charted on a map. For a man who could never navigate, these were exciting voyages.

It was such a six-week tour that Birdseye joined in the spring of 1912 as the harbors thawed and became navigable again. It is not known how he learned that Grenfell was looking for someone, but he was immediately excited by this new opportunity in Labrador.

Grenfell at age forty-seven was considerably older than Howell had been in Montana when Birdseye drove him out as too old for the job. But the Englishman seems to have had a considerable impact on Birdseye. While Grenfell never mentioned Birdseye in his 1919 autobiography, Birdseye frequently referred to Grenfell in his journals.

The older, avuncular-looking Grenfell was Birdseye's kind of man. He was a sportsman and hunter, inventive, always investigating problems and looking for solutions. Birdseye had a limited interest in religion. In Labrador he would forgo church services, even though all his neighbors were going, because with the difficulties of transportation they would use up most of his day, and he preferred "a peaceful day at home." But he would devote a small portion of that day to reading a few chapters of the Gospels. When he worked on the *Strathcona*, when the mission went ashore on Sundays to hold services with the locals, Birdseye would often go along. His strongest impressions were of the fervent manner and big voices with which the locals sang.

Though Grenfell was far more religious, his religion had a very soft touch. He said in his autobiography that he didn't like being called a missionary because it implied "someone without practical abilities and a hypocrite."

Like his mentor Moody, Grenfell had an unusually pragmatic approach to religion, one that was compatible with a secular scientific mind like Birdseye's. Grenfell wrote a story about a blind man who traveled to a shrine in Quebec to get back his sight and, accustomed to the kindness of Labradoreans, was shocked by the harsh indifference of people there. The shrine didn't restore his sight, and he only made it back to Labrador through the kindness of one stranger. Grenfell's point was that it was a mistake to look for the help of saints in heaven. It was the kindness of people on earth that mattered.

Grenfell, like Birdseye, was an adventurer forever trying out new ideas. He started the first cooperative so the locals could buy provisions at reasonable prices and not just hand over their fish and furs. He experimented with bringing in reindeer from Lapland as domestic animals that could haul goods like sled dogs.

But more interesting to Birdseye was his idea about fox farming. Grenfell, always searching for ways to get Labradoreans into cash trade and out of barter, experimented with fox farming—breeding the animals and selling the pups. It had never been done in Labrador, and it not only seemed to Grenfell to have the potential of earning a considerable amount of cash for locals but struck him as more humane than letting foxes starve to death with a leg in the steel jaw of a trap. Grenfell knew nothing about the breeding of foxes, but he was confident that it was feasible because he had seen a litter at the Washington zoo that had been born in captivity.

Grenfell established the farm in St. Anthony, a handful of scattered houses in northern Newfoundland. It had one of the harshest climates in the region, but it was a convenient location because he had a hospital there. He brought the foxes in from Labrador on his ship, and a visiting Harvard scholar recalled the frisky puppies playing on deck with dogs and people. The red foxes were particularly friendly and would run up to greet visitors. The whites and silvers were shier. But they were difficult to breed, it was hard to keep the young alive, and after a few years some foxes who had become favorite pets died of a spreading disease, and Grenfell closed down the project and turned the farm into a summer vegetable garden.

When Grenfell talked about fox farming, it caught Birdseye's interest. He had made money in furs before, and animals were something with which he was experienced. Birdseye had decided, as he later put it, that he "was not cut out for a career in pure science and wanted to get into some field where I could apply scientific knowledge to an economic opportunity."

By the time Birdseye arrived, a successful fox farm had been established in the interior of Labrador, and fox farming had become the most important industry on Prince Edward Island, with high prices being paid for live foxes.

At every port where the *Strathcona* landed, Birdseye talked to fur traders and other men involved with foxes, and he concluded that there was money to be made in trapping silver foxes and shipping them to the United States as breeding stock.

By July Birdseye was back in New York looking for an investor. According to his journal, his father wrote a letter to Harris Hammond, a wealthy contact. This is the earliest record of a link between Birdseye and the city of Gloucester, Massachusetts; the Hammond family spent part of their time there. Harris was the son of John Hays Hammond, a wealthy mining engineer who had worked in Mexico and South Africa. The elder Hammond was a close friend of the U.S. president and later Supreme Court justice William Howard Taft and had many other contacts well placed for anything the Hammond family wanted to do. When his son John Hays Hammond Jr. said he wanted to be an inventor, the father had the boy sit down with Thomas Edison and Alexander Graham Bell to talk about his ideas. John's older brother Harris was more in the mold of his father. He understood about adventures. When he was only a baby, his family had dragged him through an insurrection in Mexico. His father wrote in his autobiography, "Fortunately, he turned out to be an indestructible baby."

At the time Birdseye met the indestructible baby, he was in his thirties, only a few years older than Birdseye. He had just lost out on a major investment in Mexico because of the outbreak of what became the Mexican Revolution. He continued looking for mineral wealth and finally made his fortune at the height of the Depression in California oil fields.

Hammond perplexed Birdseye. It was difficult to get a meeting with him, and he went to his office day after day. When finally he did meet with him, he explained his idea of fox farming in broad terms. Hammond asked few questions and sought no details. He agreed to give the venture some backing and handed Birdseye a check for $750 to get him started. All Hammond said was "Keep me informed." This was the beginning of the Hammond and Birdseye Fur Company.

The next day Birdseye spent $350 on supplies and deposited the remaining $400 in a bank in St. John's, the Newfoundland capital. Among his purchases was a life insurance policy. Birdseye later said that he also got backing from Grenfell's New York connections. It would take thousands of dollars to buy enough animals to start breeding foxes.

When Birdseye got to Prince Edward Island, breeding foxes were selling for as high as $8,000, a huge sum in those days. There was a daunting amount of information and skills that Birdseye had to learn to start fox farming. He studied the farms on Prince Edward Island and drew detailed diagrams. He learned that foxes could die from drinking too much milk. They died easily, and puppies were even more fragile. They would eat cod heads, which were virtually free. "Moldy bread was good fox feed. Fish in seal oil was cheap, kept well through the winter and was a good feed. Liver tends to loosen foxes bowel. For breeding one male can serve two females although it took experience to accomplish this—that is, on the part of the breeder." Birdseye was almost never without humor.

After studying the farms' operations, he was able to find three pairs to begin with, which he bought for only $1,000 a pair. In St. John's he learned there was a proposal to ban the export of foxes, and he met with legislators to try to dissuade them from this measure. Finally he found his loophole: there was to be no ban on the export of locally raised offspring.

Birdseye then obtained a license to capture wild foxes for breeding. He also met with the fishing commissioner and other officials, went to shops, visited a seal factory. On his crossing to Labrador he met fishermen and grilled them about their techniques and their problems. He operated like a journalist constantly hunting down information.

A persistent myth is that Birdseye went to Labrador to work for the Hudson's Bay Company. This clearly is not true. Both Birdseye and Grenfell tried to interest this long-established trader of Canadian fur in the farmed-fox trade but failed to do so. The confusion comes from the fact that the Hudson's

Bay Company maintained stores in Canada that years later during World War II carried the Birds Eye line of frozen food. The company was founded by an English royal charter in 1670 and was made the sole proprietor of the Hudson Bay area. The official name was Governor and Company of Adventurers of England Trading into Hudson's Bay. In 1912 both Birdseye and Grenfell found the Hudson's Bay Company to be a very old establishment, very set in its ways, and those ways did not include fox farming, which was, given the cost of the animals and their tendency to die in captivity, a high-risk business.

Birdseye bought an abandoned outpost of the Revillon Frères, a French competitor that the Hudson's Bay Company later bought out, in Sandwich Bay, which was 250 miles up the coast from Battle Harbour, where the ships came in from Newfoundland. He seemed fond of the animals, describing some as playful, some, especially females, as gentle. They seemed almost like pets if you didn't contemplate their fate. He wrote in his journal, "Considering that the foxes are supposed to be wild, they aren't at all bashful any longer." The problem was that they were constantly dying, especially new pups.

Birdseye also had to learn many things just to survive a Labrador winter. He had to learn to dress in clothing that was heavy enough to hold in his body heat and yet light enough to keep him from sweating. Sweat would freeze and chill you, especially at night.

He had to learn about sleds and sled dogs and managing a team of them. He was impressed at his first encounter with huskies. "Here for the first time we have met the full-blood husky dog of northern Labrador," he wrote in his journal. He referred to their howling as "canine music." After listening to them through a sleepless night, he wrote that it reminded him of the noise made by a large crowd on an election night, which is as close as he ever came to a political observation.

The Labrador sled dog was close to a wolf in appearance and often in temperament. It was an extremely strong dog, fearless, and ready to attack whatever threatened, even a polar bear or a man, both of which a real wolf would shy away from. The dogs seemed impervious to cold even when covered in ice. They could save a man's life because they could always find their way even when snow blindness or fog bewildered the human driver. They average about six miles an hour but could travel two or three times that speed. Each dog has its unique personality and temperament, and a driver needed to know his dogs.

The sled, the komatik, was designed for hauling freight for scientific expeditions and for traders. It was very different from the Alaskan racing sled with the perch in the back that most people picture as a dogsled. These were eleven-foot-long platforms on long runners ideally made of light and durable black spruce, although there were many variations. There were even stories of drivers in an emergency breakdown finding whale ribs to use. The runners were slightly more than two feet apart. Packing a komatik required skill. It was extremely important that the weight be right for the terrain and that the freight be well balanced.

Birdseye traveled all over Labrador, often for days at a time, by dogsled. For instance, on January 10, 1913, he averaged almost ten miles an hour, covering sixty-five miles in six and a half hours, which is the total amount of daylight in Labrador that time of year. His journal shows him frequently gone for three, sometimes even six weeks procuring foxes or provisions. He often wrote of six-hour journeys just to visit a neighbor. Surviving the weather was challenging. It was often well below zero degrees Fahrenheit, sometimes as low as forty below zero. He later said that he regularly suffered from frostbite. "After a while you get used to it just like mosquito bites."

Provisioning a journey properly was important since there was always a possibility of spending some extra time lost in the snowy wilderness. The komatik traveler had to know what kinds of provisions to take, that salted food such as pork or cod resisted freezing and was therefore better than canned food, which would quickly freeze up and be very difficult to thaw and which also had extra inedible weight. Grenfell liked to travel with "pork buns"—bread filled with salt pork—and Birdseye picked up this practice. His first winter, though, he did not have the recipe exactly right, and so in his journal he laughed at what a rube he was and labeled his attempt "unleavened pork loaf":

> Bread with pork in it is common here—Rube forgot to take any baking powder with the barrels this trip so that the pork loaf was simply an unleavened mixture of flour, water, pieces of salt pork, and a few raisins fried brown on both sides. This bread is very soggy but each hunk is a square meal and it certainly sticks to the ribs for a long while.

More is known about Bob Birdseye in this period than at any other time in his life. He had long evenings alone, and he filled them by writing page after page to mail to his family, even though the mail-carrying ships could not get past the ice for the entire winter and much of the spring. In 1914 the ice did not melt enough for ships to arrive until July. But Bob kept writing and sent off a thick pile of pages when he had a chance.

He also kept what he called "a field journal." It is not certain exactly why he kept these journals. It may have had something to do with his training as a biologist or with working for the U.S. Biological Survey. But the style, tone, and even content of these journals, index and accounting aside, were not very different from the letters he wrote home.

He continued his relationship with the U.S. Biological Survey. At its request he was stuffing birds and other specimens to ship to Washington. An entomologist on the survey had asked for certain insects. Another wanted intestinal parasites. The Canadian Department of Agriculture was interested in plant samples. He started getting other requests. One man wanted a dozen stuffed great horned owls.

The journals were handwritten in bound hardcover doeskin notebooks with lined pages. He started the first one during his last summer in Montana and continued the practice in Labrador until his final entry in the twelfth notebook in July 1916.

The letters and notebooks tell much about Bob Birdseye, even though one of the things they show is that he did not readily talk about emotions or matters of the heart. His visits to fox farmers on Prince Edward Island reveal a man who makes very quick assessments of other people—especially observations such as "very open"—and then sucks volumes of information out of everyone he meets. Some were not open. One man he found "suspicious and secretive," and he wrote, "Most of the facts obtained were therefore wiled out of him." But that comment was after seven handwritten pages of information obtained.

These are clearly the journals of an extremely methodical man. In the early notebooks he numbered each page and used the last pages for a detailed index. In later notebooks the index was dropped, and the final pages were devoted to a detailed accounting of his expenses.

His journals also show him to be a tireless worker. After his first winter he realized that he needed help, someone to run his fox farm while he was away on trips, and he brought in a recruit from his hometown of Montclair, Perry W. Terhune. The fact that among Terhune's tasks while Birdseye was away was maintaining his journal proves that he was keeping these notes as more than just a personal record. The parts of the journal written by Terhune are in a much more careless script, with no food descriptions and no index and many days skipped or simply marked "No work done." Birdseye never had such days. He either worked on his journals or indexed them, balanced his accounts, or spent the day reading, but he always reported some profitable use of his time.

The writings to his family show that he had a self-effacing, sometimes corny, but endearing sense of humor. He frequently referred to himself as "the prodigal son," "Labrador Bob," or "the floating rib." He poked fun at his rapidly balding head, saying that Eleanor could give him a haircut with three snips behind and a rub of steel wool on top. He laughed at how lengthy his letters were, and when he learned that his brother Roger, serving in the army, was only allowed by military censors to send form postcards with certain selected phrases, he wrote to his family, suggesting that he should have a censor as well. "Wouldn't you folks welcome a rest from this long-winded journal, and a few weeks of 'I am well! I am warm!' 'I am sleepy,' 'I have frozen toe,' 'I went to Cartwright and frost burned my nose'?"

He enjoyed, was even fascinated by, children, even if as a single young man his concept of children and how to relate to them was still somewhat unevolved—largely centered on luring them with candy. In Labrador he began what became a lifelong habit of befriending children in the neighborhood. Of course, in Labrador the neighbors might be a hundred miles away. All his dogsled journeys were provisioned with candy in case he ran across any children. He wrote to his family:

The children of the families which live a long way from neighbors are always a source of great interest to me. They are for all the world like so many little fox pups. Usually when a stranger, especially a man, and especially a "skipper-man," comes into the house they hide behind the stove or under the bed or run up the ladder to the loft (if there is such a luxury), and from their hiding place peep out like so many scary little kittens. Taming wild animals has always been my hobby, and trying to make friends with these kids and tame them so they will come close enough to take a candy out of my fingers is great fun.

It is evident that personally he did not care much for candy because, uncharacteristically, he never describes what kind of candy it was. Normally, he liked to describe food, such as in his letter to his parents on rabbit: "Mostly we had 'um fried! But there's a scattered pie, and an occasional stew. One of the favorite breakfasts is fried rabbit livers, the thinnest crispiest deliciousest bacon, hot cornbread, and powerful good coffee. Doesn't that sound good?"

Despite his wanderlust, he felt close to his family and missed them. He seemed male oriented, closer to his father than his mother and to his brothers Kellogg, Henry, and Roger than his sisters. On June 20, 1915, he wrote home:

> The one thought that engrosses my mind almost constantly is H O M E. I want news—good news— of Mother and you, and of all my brothers and sisters . . . I shall probably be unable to leave Labrador by the first steamer, for various reasons. But on the second, which should leave here about July 10th, I must start out. Then in ten days more I shall be in Orange. Hooray. It makes me shout just to think of it.

All his life Birdseye was a friendly man, very accessible to all kinds of people. Is that why he changed his name to Bob? A man named Bob is easier to approach than a man named Clarence. His journals demonstrate how much he relished time spent with others, none more than Grenfell, whom he always referred to as Dr. Grenfell. "Golly it was so good to see him again," Birdseye wrote of Grenfell to his family after the Englishman visited him in the summer of 1914. "He is certainly a Prince, and it is one big life-size pleasure to meet him." Bob loved hearing his stories from his travels both around Labrador and abroad, such as his trip to Turkey.

Birdseye in those years developed a notable interest in playing poker, which is significant in a man who repeatedly claimed that the secret to his success was a willingness to gamble. His constant advice was "take chances." He once said, "All human progress is the result of gambling—the first man to use a bow and arrow was gambling his life that it would work." In Labrador he would write several pages at a time about poker games, often being facetious about the low stakes. "The stakes are enormous and a prospective player must be prepared to lose or double his patrimony in a single night. Buttons are chips. Each button costs one cent. The ante is two buttons. The record jackpot was $1.24. I am out 48 cents after four nights." On another day, June 3, 1913, he laughingly reported to his journal, "In the evening we all 6 had a game of draw poker and after two hours playing I was 2 cents out."

The writings also show that he was endlessly resourceful. He fashioned a berry harvester from a tin can and he repaired engines. Hunting, of course, had become his métier. But Birdseye also became a skilled fisherman, catching salmon and trout on the fly, jigging for cod, netting capelin.

In October 1912 he went goose hunting with Grenfell, also an avid sportsman, and Grenfell laughed at what a voracious hunter Bob was. Grenfell, who was constantly drawing, left an ink illustration in Bob's journal of the bespectacled Birdseye stomping through the marsh with the shotgun stock held high, trying to club a fleeing bird. Some have attributed this drawing to Birdseye, since it was in his journal, but it bears Grenfell's initials, WTG, and matches the style of the doctor's other drawings. Another drawing is on a separate piece of paper, and attached is a design for a fictitious commemorative medal of Birdseye bringing a goose to the crew of the *Strathcona*, who receive it on bended knees.

He had also learned medical skills from assisting Grenfell. He described an operation where Grenfell removed a second thumb from a six-month-old child's hand. "Miss Gilchrist held the kid and gave it chloroform, my aid was 'invaluable' in supporting the hand and the doctor wielded the carving knife . . . The doctor said, 'here's the thumb. Should we hold a funeral service or just chuck it into the fire?' And he walked to the stove and chucked the meat in the fire—while the mother wore a rather sickly smile."

Grenfell had enough confidence in his young friend that he left him with medicine, and apparently Birdseye did perform medical assistance for the locals during the winters. He recorded having treated some two dozen patients with complaints such as sore throats, toothaches, pain in the side, and hacking cough.

In his Labrador writing are some early signs of the inventor, the man who, faced with a problem, invents a device to solve it. One day in early 1915 he was at a house that had a tub, and he got to take a bath. He wrote that when he got back to New Jersey, he should invent a foldable rubber tub for traveling, though there is no trace of his ever having built it.

He learned everything he could about nature, about the habits and life cycles of mammals, birds, and fish. He knew that healthy salmon always swallowed capelin tail first as they swam them down, but "slinks," skinny salmon that have already spawned and returned to the salt water, would swallow them headfirst. He no doubt had opened some salmon stomachs and examined the contents to reach this conclusion. He learned most everything about the life of the salmon, the capelin, and the trout in Labrador. He wrote down his observations on the breeding habits of seabirds. He had to explore and expound on these things, for they excited him. He couldn't help himself. Once, after lecturing his family for pages about the different species of seals, their habits and coats and life cycles, he wrote, "Well, folks, this dissertation on seals is an accident, pure and simple. Honestly, I didn't mean to do it."

More than anything else he wrote about and appeared to think about food. He was virtually obsessed with the subject. Of course food was, as he frequently pointed out, a matter of survival. "Weather and Grub!" he once wrote. "Those are the two fundamental facts on which every other event hinges in this neck of the woods—or rather barrens." Once, in thanking his family for a package of preserved and canned foods that had arrived after fourteen months in transit, he said that they would be "life savers—really, perhaps."

In August 1912 he wrote down the following recipe:

Molasses Pie
For dinner today we had my first molasses pie—
and it was really mighty good. It was between two
crusts. No other flavoring than molasses was used
but Mrs. Lewis says that boiling the molasses and
adding a few spices improve the flavor of the pie.

Another recipe was offered in November:

For dinner today we had partridge—spruce and
white—pie. The five birds were placed in a roasting
pan with a little water. Potatoes turnips and onions
and boiled for a while on the top of the stove. The
dish which had already been lined with pie crust then
had two strips of crust put over the top and was baked
done in the oven. Yum! Yum!

That first year, with typical Birdseye humor, he often referred to himself in the third person as Rube:

Fried Stewed Partridge
This is Rubes favorite camp dish and is certainly the best way to cook birds I've yet run across. He quarters the birds, fries them about half done with salt pork, and then half a pan of water, finishing the cooking with a dish over the top of the pan. This gives moist meat and a lot of delicious gravy to be used instead of butter. Some times he mixes up some flour and pork and stews it with the partridges.

He often mentioned in either his journal or his letters home what he ate for breakfast or dinner. A typical entry was "After an early breakfast of bread and tea and salt fish, we took Fred Brown and Uncle Tom aboard and went to Dove Brook." Although Birdseye did not provide a great deal of information about himself, very few historical figures have offered so much on what they were eating.

On October 13, 1914, he sat down at his fox farm and began a letter to his family:

Well, folks, having just disposed of some toast and cocoa after a ten-mile ante breakfast walk from Cartwright in a roundabout way to Muddy Bay, I feel in good humor to inflict another chapter of this letter.

Some of his food descriptions were ecstatic. He wrote to his family:

Oh you poor half-starved city folks! Can you stretch your atrophied imaginations to see and smell that great big platter with three roasted ducklings lined all 'round with stuffed snipe, all browned to a turn?

And another time:

Today's big event was the goose dinner! As I sat at this table all the morning writing the odor of roast goose wafted in to me and gave me such an appetite that I had to eat twice as much as any of the others to satisfy it, I wish some of you folks could have sat down to dinner with us today. There was a big brown roast goose, a dish each of boiled spuds and boiled turnips, and a big bowl of boiled cabbage leaves.

Laughing at himself, he wrote to his family on May 10, 1915, "Every page has to contain *something* about food. So here are the two items of principal interest today." They were that he had "begged" a can of real butter from a neighbor and that he was able to get some fresh seal meat.

He often joked about his food obsession to his family. He once wrote toward the end of a very long letter, "Well! Well! I've taken time out for a mug-up of mince pie (honest, and mighty good too) and now feel equal to a few more sentences. Me thinks I can stand more of this than you folks can."

And as was always his style, the more exotic the food, the more enthusiastically he received it. His palate, like his mind, was endlessly curious. He wrote to his parents that a porcupine he had eaten was "unexpectedly tender, in spite of the beasts age and sex." He described as the "pièce de résistance" of "one of the most scrumptious meals I ever ate" a lynx that had marinated for an entire month in sherry and was then stewed and served with a sauce made from the marinade. He said he ate polar bear and professed a particular fondness for the front half of a skunk.

He loved seal meat, especially that of the ringed seal, which he correctly identified as *Phoca hispida*, except for old males. He informed his family, " 'orned Howl (horned-owl) for Sunday dinner—does that sound good? Well, it *was* good, no matter how it sounds." He also ate beaver and a wide variety of birds, including hawks. He explained to his family that he was able to eat fresh meat every day by having "no fool prejudices" and eating "anything that tastes good." He was so enthusiastic about these foods that he threatened to can some of them and bring them back to New Jersey "so you folks could sample them."

He even studied local books to learn about food traditions and was very excited when a book at the Grenfell Mission explained the English origin of the Labradoreans' tradition of eating pea soup on Saturday nights.

As he approached his thirties, Bob was a man who had not yet decided what to do with himself. He did not yet think of himself as an inventor. He certainly dreamed of being more than just a memorable character, though he wanted that as well, posing for a portrait in sealskins that bears a striking resemblance to a well-known photograph of Robert Peary. But on closer examination, inside the sealskin hood is not the solemn-faced admiral with the big mustache but a clean-shaven man with thick glasses and a huge smile, incapable of concealing how funny he thought he looked in the outfit.

Despite his truncated education, he thought of himself as a biologist. Perhaps that was why he continued to keep a field journal. But after his decision in 1912 that he did not want to be a theoretical scientist, he focused on commercial opportunities. Running out of money to finish college may have traumatized him. He came from a family that to an unusual extent for the nineteenth century believed in college education, even for the girls. All his life he found a variety of ways of referring to his lack of education, often talking about the things he didn't know.

In Labrador he was still searching for his career, but he already seemed drawn to the idea of having numerous careers, something he took great pride in later. Toward the end of his life he liked to tell the story about a young man asking him if he had it to do over, would he choose the same occupation, and his answer was "Which occupation?"

In Labrador he was still very interested in photography. When he first arrived, after less than a week there, he shipped back his first six rolls of film. The locals loved to pose for pictures, which they called being sketched off, as in "Would you sketch off me?" Would you take my picture? His first year there he seemed to have had the idea of a book of photographs of Labrador. In July 1913, when in New York, he dropped off a proposal with photos at Scribner's. He never commented on the response, but no book deal was forthcoming. On September 5, back in St. John's, Newfoundland, Birdseye sold twenty-five prints to a publisher to be used in a picture book on Labrador. He sold the prints for $1 each and retained all future rights. Bob was always astute about his business agreements.

He also seemed interested in writing, which may have been influenced by Grenfell, who wrote and published engaging tales of the people he met in Labrador. Then, too, Birdseye's father and his sister Miriam were regularly publishing books. In November 1913, *Outing*, a sportsmen's magazine from New York, published an article by Birdseye titled "Camping in a Labrador Snow-Hole," about an incident that had happened a year earlier. The style was not unlike that of his journals, regularly stopping to tell you what he ate. "We had a cup of tea and stewed seal meat at Tom Paliser's house," he explained in the first paragraph. Later in the day they made a pudding of rice and raisins. They were traveling by dogsled, and he had a hired driver. The next morning they had salt cod, bread and molasses, and tea for breakfast. He described how difficult it was to manage a dog team and to travel by sled and how they were trapped in a blizzard for three days with nothing to eat but raisins, prunes, rice, and the candy he always carried to give out to children. The experience nearly put an early end to Bob Birdseye, and in fact five of his dogs did not survive. The writing did not show much promise, nor was a flair for storytelling exhibited, and it did not even have the charm of a Grenfell story. But it was illustrated with his photographs, the admiral-in-sealskins portrait, and shots of his sled and dog team.

A month later *Outing* came out with his article titled "The Truth About Fox Farming." This was more like the Birdseye who would be known in later years—everything you could imagine about fox farming, complete with his diagrams and of course a few photographs of his and other fox farms. His photography was without artifice, intended purely to document. It was amateurish, but at this time there were not many amateur photographers, so he was a pioneer.

While the fox piece seemed the type of writing for which Birdseye was most suited, he was still experimenting. And he was not afraid to try new things. In 1915, *Outing* published Bob's first work of fiction, destined to be his only such work, "Hard Luck on the Labrador." Birdseye still liked fiction and often read novels and detective stories along with nonfiction. This very short piece does not have a lot of story. A man in Labrador is out hunting seals. He misses one, then he hits another but only wounds it, and the seal escapes. The story is a first-person narrative written in an attempt at Labrador dialect that is barely readable: "So after we'd et our hard-bread and drinked our tea we started back home in d' flat." Yes, he didn't forget to tell us what his character ate. He could also not resist being informative, so when the narrator gave local names for wildlife, Birdseye added the correct name in parentheses.

In late June 1913, with great excitement, Birdseye went back for a visit to Montclair. He spent a lot of time with his brother Kellogg and some time with his father, who accompanied him to visit Hammond. Hammond seemed very pleased with Birdseye's report and proposed forming a small company and even issuing stock. Birdseye would then draw a generous salary. Bob, for his part, neglected to mention that his entire Sandwich Bay operation, in a place aptly named Muddy Bay, had been obliterated by spring flooding. Hammond gave him money not only to buy supplies but also to buy a boat.

Somewhere that summer between lunches with Kellogg, visits to Kellogg's office, visits to his father's office, meetings with Hammond and meetings with furriers, showing his photography book proposal to Scribner's, and even a visit to Amherst, Bob probably managed to see Eleanor in Washington or New York. But he mentioned nothing of this in his journal, and by August 28 he was back in Newfoundland preparing for Labrador's early winter.

The following summer of 1914 he visited home again. It seemed almost certain that summer that Europe was about to go to war. This had huge implications for Birdseye's fox business since not only did he get his cash flow from British investors—possibly a Grenfell connection—but also the primary market for fox fur was Europe. He wrote in his journal, "Because of the war possibly causing some of the pledged English underscriptions to be withdrawn we might get into serious straights [*sic*], for want of capital."

Birdseye scrambled in New York. He struggled to catch Hammond in between the young financier's regular summer trips to Gloucester, and met with his father and brother Kellogg in Amherst, and met in New York with his brother Henry, who was also an investor in the company. He tried and failed to get the Hudson's Bay Company to buy him out.

In any event he didn't have much to sell. After two years in which he traveled, by his estimate, five thousand miles by dogsled gathering foxes, his business had been closed by the Newfoundland government, which finally banned the export of live foxes from Labrador and Newfoundland. This came at a time when the conventional wisdom was that the high-quality fur business was dying. American women seldom wore such furs. It was a European look, and Europe was about to be consumed with war. The principal fur centers, London and Leipzig, were both shutting down. The Hudson's Bay Company and Revillon Frères stopped buying furs.

The fur market in New York was almost as dismal. Top-quality ermines were selling for fifteen cents each. But Bob saw opportunity. The United States, the one developed country that was not at war, was prospering. Surely this new prosperity would be reflected in fashionable women in furs, especially with pelts available for bargain prices. This was Birdseye's kind of gamble. And there was one thing that could be seen over and over again at critical moments: Bob Birdseye was a great salesman. His obvious intelligence, his ability to articulate his ideas, and his contagious enthusiasm almost always carried him through. He convinced a New York furrier, who staked him to $8,000.

Back in Labrador he began killing his foxes, freezing them by packing them in snow, and sending them to be skinned. He also started buying quality pelts wherever he could find them for cut-rate prices. He traveled thousands of miles by dogsled once again, buying furs, offering low prices but in cash. The big fur buyers rarely offered cash, so he was able to make very good deals. Grenfell, who had long fought to get companies to pay cash to the locals, was able to raise capital for Birdseye also. On one trip, recorded on March 27, he had bought pelts of three red foxes, eight silver, sixty-one marten, five lynx, twelve mink, and forty-two weasels for $1,838. Most of the value was in the silver foxes. Labradoreans were selling their furs for historically low prices. He wrote of how he bought one top-quality silver fox that a year before would have been worth $4,000 for $375.

"The people were so hard up they had to sell their fur,"

Birdseye wrote to his parents. "And they certainly were glad to have me buy it, for no one else was doing so." In fact, the people were so hard up that soon he stopped his buying trips because they brought furs to sell to him. He wrote that they would take "any price I see fit to give." He did express some remorse for exploiting poor and desperate people, even commenting on how badly dressed their children were with their leaky boots. But he assuaged his conscience, perhaps a little too easily, by pointing out that he was still paying "twice what the Hudson Bay Company is giving." By the end of 1914 he had cornered the Labrador fur market and cleared $6,000 profit, a considerable sum at the time.

In the meantime Bob's parents were in complete despair. Their youngest son, Roger, had secretly gone to Canada and joined the Eighth Royal Rifles. The *New York Times* identified his outfit as the "First Canadian Contingent." The Royal Rifles was a Canadian contingent of the British army, originally formed in the eighteenth century to raise soldiers in the American colonies to fight the French in Quebec. They were to be an American regiment fighting for the empire. Now they were going to fight in Europe, and Roger Birdseye was going with them. "He was brave, if foolish, to do that," Bob wrote in his journal. "If he lives through unmaimed he will be the better for it. But it's going to be mighty hard on mother and father."

It was hard on Roger too. He had volunteered in August 1914 as the war began and was the only American in a completely Quebecer battalion. He rapidly rose in rank from private to lieutenant and according to Birdseye family history was the only American to become an officer in the British army without relinquishing his U.S. citizenship. The Second Battle of Ypres, a town in Belgium, was actually a series of battles and one of the most infamous World War I slaughters. It was the first time the Germans used poison gas on the western front. In just one of these battles one thousand Canadians were killed and almost five thousand more wounded out of a force of ten thousand. The remainder took 75 percent casualties in the next action. Roger was awarded the Distinguished Conduct Medal for carrying a wounded soldier through heavy fire. He was one of the few unwounded soldiers left in the battle when he himself was finally wounded. The *New York Times* reported that most of the men and officers in his outfit were lost and that he was the last casualty in his unit. But he survived and in 1919 married Effie May Dixon, a Canadian nurse he had met on his transport ship when he was first sent to France in 1914. He later moved to Arizona and became a publicity agent.

In July 1915 Bob left Labrador, and from St. John's he wired his parents to say he was coming home. According to his journal, he also wired "EG." This was the first time he ever mentioned Eleanor Gannett in his journal, and all it said was that he "wired EG." Later in life he would explain that the substantial earnings from fur that year made him feel ready to marry the girl he loved. Conventional wisdom sometimes claims that once people get married and have children and become weighted with responsibilities, they become less creative. But sometimes having people to protect and look after, thinking about people other than yourself, gives you your best ideas. Though it took him some time to realize it, that is exactly what happened to Bob Birdseye.

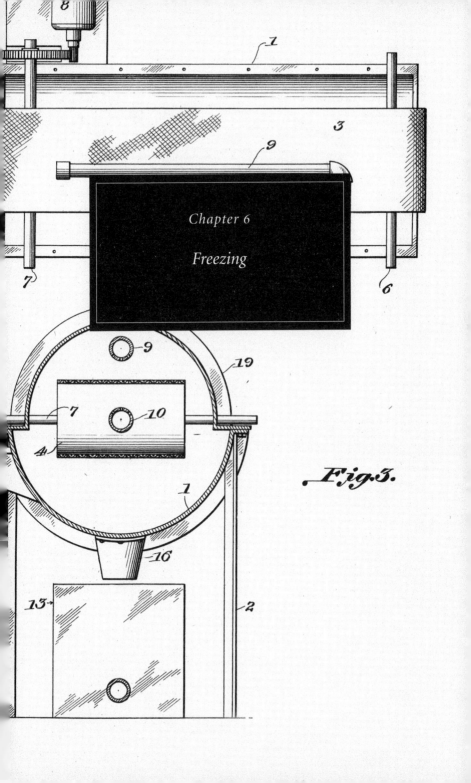

Chapter 6

Freezing

Fig.3.

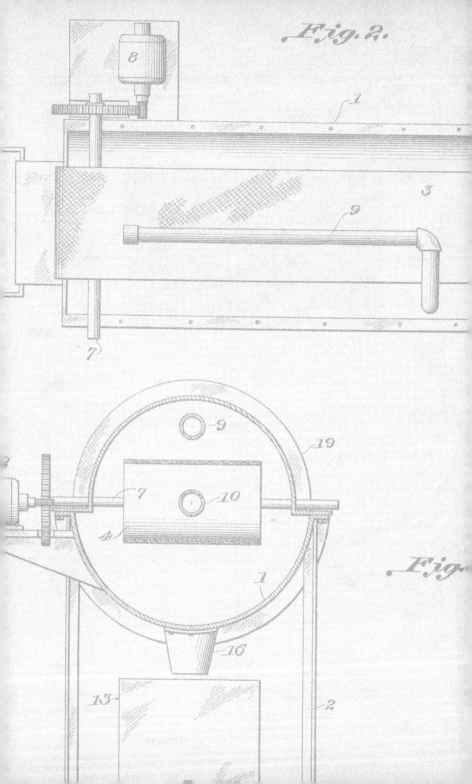

Fig.2.

Fig.

On July 25, 1915, Clarence Birdseye was back in New Jersey, and that night he sat down with his older brother Kellogg to discuss business. And he told him that by the way, he was engaged to marry the one he kept referring to in his journal as EG. This EG business is curious. It was not that he identified everybody by initials. He writes not that he told "KB" of his engagement to EG but that he told *Kellogg* about EG. It was almost as though he didn't want her to be identified in his journal.

He had a meeting at Kellogg's office with Louis Underwood, a General Electric engineer and executive who was interested in Labrador and was probably a cousin. He made a number of mentions of shopping for a diamond but didn't say why. Feeling financially secure because his furs had been assessed and his fur company showed an estimated profit for the year of between $3,000 and $4,000, he finally chose a diamond and wrote that he was having it set as an engagement ring.

For almost an entire month there were no entries in his journal. Then, on August 29, he returned with Eleanor by car from Berlin, New York. He described the hotel in Berlin, the Taconic Inn owned by T. D. Taylor, and wrote that rooms were $20. He described various people on the property and who their families were. Birdseye was always interested in people. A reader of these journals would know more about the young couple that gave them a tour of a nearby cave than he or she would learn about EG. He included in the journal the torn upper half of two pieces of stationery that were the letterhead with a picture of the Taconic Inn, which boasted electric lights and steam heat. Saving these and pressing them in his journal were, for Birdseye, rare sentimental gestures. But why not? Though he never mentioned it, this was his honeymoon. He also didn't mention that a week earlier, on August 21, they were married. But at least he now started referring to her in the journal as Eleanor rather than EG.

Life had now changed. Suddenly the preparations for returning to Labrador included not only the usual shopping for shotguns, ammunition, and equipment but a week of daily shopping at Wanamaker's. This was one of America's first department stores, founded in Philadelphia in 1876. The founder, by coincidence, was also a fan of Dwight Moody and had once let his Philadelphia store serve as a meeting place for the popular evangelist. In 1915, New York's Wanamaker's was at its height, known for its numerous and helpful staff and for inventing thematic sales such as "the white sale." It had its own wireless terminal and in 1912 was one of the first places in New York to receive news of the sinking *Titanic*.

In Wanamaker's day after day, Eleanor and Bob's sister Miriam bought bedroom furniture and dining room furniture and a great deal of aluminum and white agate kitchenware—all to be shipped to Labrador. Bob did take the time to sell two pairs of snowshoes he had brought from Labrador to Abercrombie and Fitch for $10 each, a neat little profit. New York was no longer just about business. They all went to the American Museum of Natural History together, an old friend for Bob, and then to their first movie, D. W. Griffith's pioneering but shamelessly racist *Birth of a Nation*, though there is no record of what they thought of it.

With what we now know about Clarence Birdseye, Miriam would seem the sibling with whom he had the most in common and not his businessmen brothers. Miriam was also fascinated by food and at the time was lecturing on nutrition, particularly about meat, at Cornell University. But it seems she only became closer to Clarence with the arrival of Eleanor. They were both college-educated women and by coincidence would both later write about gardening—Eleanor on wildflowers and Miriam on herbs with the writer Leonie de Sounin, with whom Miriam lived and gardened for many years.

Then Bob and Eleanor were off to Labrador by the usual route—train to Boston, train to Maine, boat to Halifax, train to Sydney, boat to the tiny seaport of Port aux Basques, train across the rockbound, barely populated wild interior to St. John's. It is the same route that is traveled today except that the train from Port aux Basques to St. John's, which used to be called the Newfie Bullet, no longer runs, replaced by a usually empty highway.

They had to wait two weeks at a hotel in St. John's for the boat to Labrador. Compared with anything in Labrador, St. John's was a big city, but in fact it was just a rugged fishing town, not even as big as Halifax. It dated back to before 1620, but its bustling waterfront had more of a sense of commerce than history. Once the harbors thawed, the town became a tower of Babel as ships came in with French-speaking Bretons, Portuguese, and all the other long-range fishermen of Europe who came to fish cod in the Grand Banks and put in to St. John's for supplies and repairs. The harbor was packed with the fat-hulled, square-rigged barks of the European fisheries, the sleek fore-and-aft-rigged schooners that fished out of Gloucester and Nova Scotia, and newer engine-powered vessels hauling freight and passengers.

In October they got to Labrador and started building a new house. Bob would name the house Wyndiecote, the old Long Island name that to him meant home. Birdseye wrote that he was spending all his time on this project. Perched on a ledge of rocks over Muddy Bay, it was a simple house, though luxurious by the standards of the shacks that most people in the area had for homes. Built by a local trapper, Charles Bird, it was set on a platform that was reached by way of a long stairway up the rocks. The large living room was surrounded by windows looking out on the sky, sea, and more distant woods. It was a one-story house with only three rooms, but solid. It still stands, though its location has been changed. Among its many owners have been the International Grenfell Organization and the Hudson's Bay Company. It has been preserved not for its historic importance but because houses are scarce in Labrador, and you don't throw away a solid one. The Birdeyes finally moved in on December 9, which happened to be Bob's twenty-ninth birthday. He mentioned in his journal that it was his birthday, the only time in five years that he ever mentioned the occasion. Things were changing.

From Newfoundland, Eleanor wrote to Bob's mother that she was perturbed by the inefficiency of the people. "No one in any line of business thinks of starting the day's work before nine," she complained. She didn't like the way the midday meal used up too much of the working day and thought that all of these bad work habits explained why they were still utterly dependent on England for all their goods, which, she added, made the local shopping "most unsatisfactory."

They hired a live-in maid, not uncommon at the time, back where they came from. Bob now took Sundays off, and they read and went for walks in snowshoes in –14-degree temperatures.

For all her straitlaced sense of the right way to do things, Eleanor was ready for the adventure. She traveled with Bob on long journeys with a nine-dog sled, setting traps in subzero weather. Years later she liked to amuse her children with a story of the time she fell off the back of a komatik and Bob, up front managing the dog team, didn't notice for about ten minutes, in which he covered about a mile. He looked back and saw no Eleanor. So he turned around and found her. She always laughed when she told the story, but there must have been a dark twenty minutes or more when she was abandoned in the Labrador winter with no snowshoes or provisions.

Eleanor adapted quickly to life on the frontier. They did target practice together with .22 Colt revolvers, which they also used to shoot rabbits that were caught in their traps. At first, though she did well with targets, Eleanor could never hit the trapped rabbits. But in 1917, Birdseye wrote home that Eleanor could make a head shot at fifteen yards " 'most every time," which, given the size of a rabbit head and the accuracy of a handgun, was fairly good shooting.

Eleanor gradually found her place with the Hammond and Birdseye Fur Company, doing paperwork and looking after foxes, which apparently they were still breeding, though they could export them only if they killed them. She also undertook to establish a darkroom in the new house so that they could develop the film and make prints from Bob's photography.

In May fox pups were born. The journal and apparently their lives became consumed with caring for one struggling pup. Bob writes in his journal when the pup gets his first teeth and when he has "a copious movement" due to the use of castor oil. Finally, on May 18, 1916, Birdseye reported that the pup died at 6:00 a.m. of acute indigestion.

At the same time hundreds of thousands of Frenchmen and Germans were dying in a place in France called Verdun. But Bob and Eleanor would not hear about that until the harbor thawed in another month. Everyone in Labrador was eager for scarce news of the war. The dominion had given more than its share to the war effort. An entire Newfoundland regiment was slaughtered on the first day of the Battle of the Somme in 1916. Most war news trickled in as rumors. Typically, in a 1915 letter to his family, Birdseye commented that he had heard that "Belgium had wiped out a detachment of German soldiers. I certainly hope that is the case!"

In the meantime they had foxes to worry about.

Once the harbors thawed, they went back to New York. Eleanor was more than six months pregnant, but this was the earliest they could leave. Her pregnancy had not taken up nearly the space in the journal that the fox pup had. Once in New York, Birdseye simply noted, "Shortly after breakfast E and I went to Dr. Stern's office and made arrangements. I gave him carte blanche in making hospital and all other necessary arrangements." That taken care of, Bob tried to work on business in New York. After numerous attempts to meet with Hammond, who seemed always to be in Gloucester, Birdseye was informed, "Mr. Hammond is no longer interested in Hammond and Birdseye." Bob's father tried to set him up with other prospective investors, and apparently something worked because he went back to Labrador, where he awaited the return of Eleanor and their new baby.

On September 6, 1916, their son Kellogg was born, and Eleanor returned to Labrador with the five-week-old baby. The ship she was to sail on never arrived in Newfoundland because a German U-boat had sunk it. Finally, she managed to get on the last boat for Labrador before ice shut the ports for the winter. The ship even sailed through a storm to arrive in Labrador before it iced up. To raise a baby in a subarctic wilderness, with the closest medical facility, Grenfell's hospital, 250 miles away by dogsled in Battle Harbour, seemed a risky decision. But Eleanor and Bob were not troubled by it. Years later the family would say, "But what if Kellogg got sick?" and Eleanor would smile softly, shrug, and say, "But he never did."

To the biographer the first obvious change now that Birdseye was a father was that he stopped writing. He dropped his journals altogether and wrote far less to his family. The family would still hear from him whenever there was a mail ship, but the packets were much lighter. This suggests that one motive for the writing, which was mostly done at night, had been to stave off loneliness. He urged his parents to keep writing back "the same number of good letters that you always sent me when I was a poor forlorn bachelor." Though he was always understated, it was clear how much Bob enjoyed having Eleanor with him. He described her as "the only genuine Washingtonian north of the forty-nine." Eleanor began writing to the Birdseyes too, addressing them as Mother Bee and Father Bee, a habit their daughters-in-law picked up for addressing Eleanor and Bob.

Typical of Birdseye, he did not worry a great deal about the proximity of a hospital, but he was concerned that Kellogg, whom he always referred to when writing to his family as "Sonny," ate well. Ringing in his ears was Grenfell's constant admonition to the people of Labrador that poor diet was causing widespread anemia, dyspepsia, beriberi, and scurvy. Fruits and vegetables had to be brought back from trips to Newfoundland. Livestock for meat or milk could not be kept, because the dogs would kill them, and they had to keep the dogs. Sled dogs, the only means of transportation, were not optional. Birdseye thought a lot about preservation because of the long winters he had to provision for. These are two recipes from his journals:

Dried Trout

Trout are slit up the back letting them stand overnight in a little salt, washing out the brine next morning. And partially drying them in the sun. By smoking them as are salmon, trout can be kept all winter.

Dried Capelin

Spread out on bare rocks, roofs, and flakes, everywhere are capelin drying. They are simply left overnight in their own juice (½ gal salt to 1 lb fish) raised out next morning. And tossed on the handiest surface to dry. They are used either for white folks or dogs during the winter.

He also had to learn how to preserve food to feed the foxes. He used partly dried capelin that he barreled with one-sixth their bulk in seal oil and found that this formula preserved the fish so that they would keep a very long time and made good fox food. Birdseye estimated that with this feed it cost him between $5 and $6 to feed a pair of foxes for a year. This was a considerable improvement over the Prince Edward Island breeders, who spent about $50 a year feeding each pair of foxes.

Fish was the cheapest and most plentiful food. One of the summer tasks was to preserve salted fish. In the winter fish was frozen by a technique called snow packing in which food was buried in a barrel of snow and left outdoors. Since the temperature was below freezing all winter, it remained frozen until unpacked for use.

Birdseye asked himself many questions about food and survival in the subarctic. Why, he wondered, did people in Labrador eat lean food in the summer but a tremendous amount of fat in the winter? The ultimate winter survival dish was something he called bruise, which is sometimes known as brewis, a combination of dried and salted food mixed with a tremendous amount of fat. Usually it was salt cod, hard-tack, flour, and water, baked hard and mixed with cubed salt pork, and then boiled and served like a hash with huge globs of melted pork fat. Bowls of melted fat were often served on the table to spoon onto food. Birdseye laughed when he heard a host say, "Have some more grease on your bruise," but everyone then took a few spoonfuls. It was a Sunday morning breakfast favorite. He remembered that people also ate a great deal of grease in the Southwest, where it was hot in the summer. They would open a can of corn and eat it with pork fat.

Birdseye reflected on such phenomena. He asked his parents in a letter why people in hot and cold climates eat more fat than those living in a temperate zone. But it may not have been related to temperature. It might have been that people living in wildernesses use a lot of energy in their lives, and since food is hard to come by, fat is sometimes both filling and available.

The constant diet of grease and preserved and conserved food gave him a longing for food that tasted fresh. "Good lord," he once wrote to his family, "how fine gull gravy tastes when one hadn't had anything fresh for a long time." It was not only Birdseye. He told his parents that whenever someone encountered other people in Labrador, the first thing asked was whether they had seen any fresh food where they came from. He often described the excitement and relish with which neighbors would gather and sit down for a meal because they had gotten some fresh food. When he bought foxes, he always tried also to buy beavers, lynx, marten, otters—whatever he could get. He wanted them with the fur intact so that he could stuff them and sell them as specimens. But he also wanted to eat the fresh meat. Everyone in Labrador was craving fresh food in the winter. Although Birdseye loved irony, he was being absolutely serious when he referred to a meal with fresh food as "Today's big event."

This hunger for the taste of freshness had a lasting effect on Birdseye. It is easy with the wisdom of hindsight to say that naturally a young man looking for commercial opportunities who had an obsession with food and a passion for the taste of freshness would make his fortune developing better-preserved food, and that, coming from Labrador, where everything froze, he would work with frozen food.

But that is not what happened. His food concerns were much more immediate. He was not thinking of ambitious plans to launch a new food industry. He was just trying to make sure his young son and his wife ate well. Making that task even more difficult, Bob was reluctant to travel and leave Eleanor and Kellogg alone. He wrote in a 1917 letter, "Pulling teeth would have been a mild process compared to running off and leaving E. and Sonny."

The Birdseyes brought live hens with them from Newfoundland and also a large supply of fresh potatoes, turnips, beets, carrots, parsnips, cabbage, onions, apples, and grapes. Also thirty dozen eggs because they did not anticipate the Rhode Island Red hens would start laying until the spring. They also shipped with them from Newfoundland two whole beef hindquarters and a whole lamb.

"So you see that our larder is going to be well-stocked," Bob wrote home to New Jersey, "and we needn't fear scurvy or rickets, or pip, or beriberi or any of those little ailments which come of a too salt diet!"

All he had to do now was figure out how to keep this trove in reasonably fresh condition throughout a long winter. In the event of failure, they also shipped a large supply of canned fruits and vegetables. The grapes froze on the way to Sandwich Bay, and the eggs spoiled quickly and could not even be fed to the foxes. But by the following spring the hens were producing. Birdseye froze several hundred partridge in the fall. Another reason why freezing, snow packing in barrels, became more important with the arrival of Eleanor was that she disliked gamy-tasting meat, so a wild goose or a partridge kept unfrozen even a few days was too ripe for her taste.

All of this was driving Birdseye's lively intellect to ponder on the science of freezing. He spent endless time reflecting on things that would have barely registered in most minds.

In December 1914, when he was still single, he made a simple observation on something that had plagued his curiosity for years. He wrote to his father:

Practically every morning throughout the winter the water in my pitcher is frozen—often so hard that it has to be thawed out with hot water. So a few mornings ago, after a cold night, I was much surprised to find, upon thrusting a hand into the pitcher to find it filled with water instead of ice. When, however I poured some of the water into my bowl and some more into a glass, and then scooped up a handful I found that the bowl was full of a spongy mass of ice crystals; and the same formation had taken place in the glass and the water pitcher—yet a few seconds before there had been no sign of ice in the water. Evidently the water had been in a state of equilibrium—at the freezing point, and all ready to congeal, but needing some little stirring up to start the crystallization. I seem to remember seeing the same thing done in a physics lab experiment, but certainly never ran across it before "in nature." Did you?

Some days later he commented on the phenomenon to his family again, adding, "Some of you physics-sharks please give me an explanation of this happening." This is the earliest record of Birdseye contemplating the science of freezing and the laws of crystallization.

Freezing was not new. There had been frozen food available all of his life. But when it thawed, it was mushy and less appealing than even canned food. Frozen food was a last resort. No one wanted to eat frozen food. But to Birdseye's surprise the frozen food in Labrador was not unpleasant. In fact, in his judgment, it tasted just like fresh food. What accounted for the difference? The Inuit had traditionally enjoyed this high-quality frozen food. They fished in holes in the ice and pulled out a trout, and it instantly froze in the thirty-below-zero air. When they cooked it, it tasted like fresh fish. In fact, sometimes the Inuit would put the frozen fish in water and thaw it, and the fish would start swimming in the water, still alive. According to Birdseye, some fish were still alive after being frozen for months. Birdseye spent years trying to understand the mystery of the live frozen fish, but he did learn to ice fish, instantly freeze the fish in the air, store it outside in the cold, and thaw it in water when ready to cook it. And he learned to snow pack fresh meat. Frozen game was so fresh when it thawed that it did not taste in the least aged, and even Eleanor liked the taste.

He noticed that the meat and fish were not as good when frozen in the early or late winter, and he wondered why. He would cut paper-thin slices of the frozen food and see that the food frozen early and late in the winter did not have the same texture as the food frozen in the dead of winter. He could see the difference. The spring and fall food had a grainier texture and leaked juiced when thawed.

It was clear that in the dead of winter, when the air was thirty below or even colder, the food froze instantly whereas in warmer weather it took longer to freeze. This was not hard for Birdseye to understand when he thought about it. It had to do with the science of crystallization, of which everyone in Labrador had some knowledge because preserving food with salt was a way of life there. Salting food and freezing food are opposite processes. Freezing needs small crystals, and salting needs large ones for the best results. For centuries sea salt had been in great demand in places like Labrador with a great deal of fish but not a sunny enough climate for solar-evaporated sea salt. The only economically viable way to produce sea salt is by evaporating seawater in the sun. The cost of fuel made cooking it down economically unworkable. The fish required sea salt not because, like the fish, it came from the sea but because of its large crystals. Solar evaporation is a very slow way to make salt, and so the crystals are very large. This is the rule with any kind of crystallization: the more slowly the crystals form, the larger they are.

Birdseye looked at the inferior frozen food with its grainy texture and leaking juices and realized that the ice crystals were too big. If food is frozen too slowly, the larger ice crystals will damage the cellular structure, even break down cells. Everyone knew that the height of winter was the best freezing season. Birdseye figured out why.

He started experimenting with vegetables. When the weather turned very cold, he took a large barrel and put an inch of seawater at the bottom. Then he put a thin layer of cabbage leaves. He had bought a lot of cabbage in Newfoundland and had been storing it in the house until the winter got really cold. Once this was frozen, he added another layer of seawater and another layer of cabbage. He repeated this until he had a full barrel of cabbage. When he wanted some cabbage for the family, he would lob off a chunk with an ax and cook it, and he found that it tasted exactly like cooked fresh cabbage.

Not all attempts at freezing food were successful. He butchered a caribou and froze it inside blocks of ice, only to find out months later that it had not been frozen cold enough; the natural salt in the animal blood had melted the interior of the blocks and the meat had rotted. Birdseye liked to joke years later that this was the original "Birdseye Frosted Meat."

On April 6, 1917, the United States, despite President Woodrow Wilson's election campaign promise not to, entered World War I. Birdseye always said that he returned to the United States because of the war. But he did not clarify why. Others said the motivation was patriotism, but what patriotic act was he returning to perform? He did register for the draft from the new Birdseye family home in Englewood, New Jersey, but he applied for an exemption as the father of an infant, and the military was not interested in a thirty-year-old father.

It may have just felt like the right time to go home. He had been Buffalo Bill and Admiral Peary, and he was planning on more adventures. He was married and a father, and he intended to have more children. Their second child, Ruth, was born in 1918. It was time for a different kind of life.

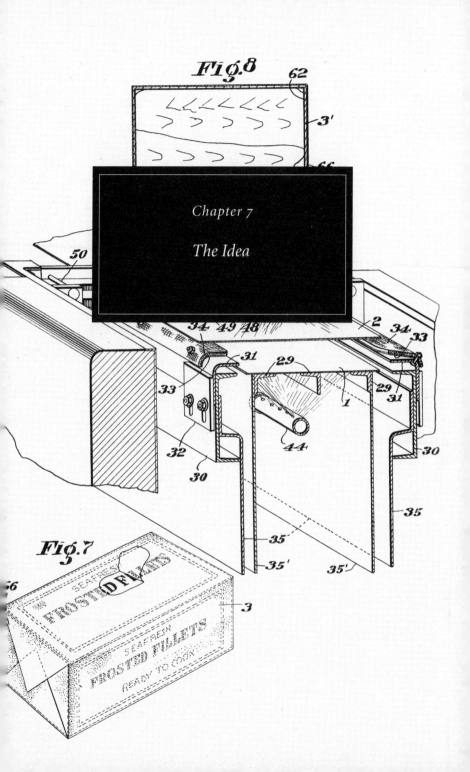

Fig. 8

62

3'

66

Chapter 7

The Idea

50

34 49 48

2 34 33

31 29

33 29 31

32

30

1 44

30

30 35

35

35'

35'

Fig. 7

66

SEAFRSH FROSTED FILLETS

3

SEAFRSH FROSTED FILLETS READY TO COOK

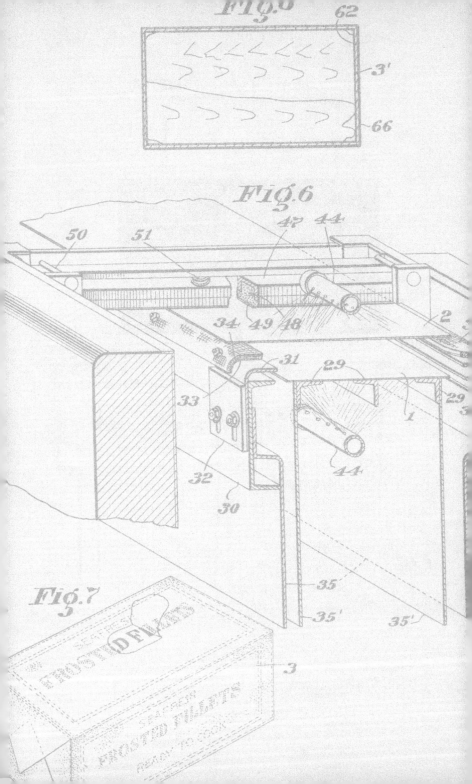

Fig.6

62

3'

66

Fig.6

50   51   47   44

34   49   48   2

31   29   29

33

32   30   1   44

35

35'   35'

Fig.7

SEAFRESH
FROSTED FILLETS   3

SEAFRESH
FROSTED FILLETS
READY TO COOK

Bob and Eleanor Birdseye had left Labrador, and they were no longer preoccupied with the issues of subarctic survival. They were no longer confronted with problems such as fresh food supply. Birdseye had no reason to think about freezing food anymore. He was now living in a world in which frozen food could be completely avoided, which was exactly what most people wanted to do. He had no more reason to think of the commercial possibilities of freezing than he did about snowshoes or dogsleds. He later said of what he had learned about freezing, "I tucked this knowledge away in my subconscious mind, but its commercial possibilities did not dawn on me at that time."

There was something else on Bob's mind. In May 1917 Bob's father, Clarence, and older brother Kellogg were arrested and charged with conspiracy to commit fraud. Whether or not this was a factor in Bob's return to the United States is not known. Bob had seldom taken an important step without consulting both his father and his brother Kellogg. They had been his regular advisers in his years in the Labrador fur trade and the two family members most often mentioned in his Labrador writings. Bob and Eleanor had named their first son after Kellogg.

The arrest got some press attention. On May 5, the *Standard*, a weekly that covered the insurance industry, reported the allegations in great detail and said, "The story is one of the most extraordinary in the history of life insurance." Another newspaper covering the industry, the *Insurance Press*, on May 9, under the headline "The Looting of the Pittsburgh Life and Trust," wrote, "A narrative of astounding effrontery and rascality."

Clarence senior and Kellogg had become involved in a life insurance company, Pittsburgh Life and Trust, of which Kellogg became treasurer. Clarence managed to get control of the company and establish his own board of directors, who let him dip into their safe-deposit box and take securities, which he sold and pocketed the proceeds of, amounting to some $1.9 million, worth almost $22 million today. But this was only part of a far more complicated plot that involved luring in a lumber company and numerous other players to cover up what the prosecution termed looting. They were gutting the company, and New York and Pennsylvania authorities stepped in to protect the insured. Accounts of the trial make clear that this was not a small oversight or a slight brush with the law but an elaborate conspiracy to steal a fortune. It appeared that Clarence Birdseye Sr., a leading authority on New York State law whose books were studied by law students and lawyers, was the ringleader in a scheme involving lying to and deceiving a considerable number of business associates, many of whom testified against him.

Bob and Eleanor and young Kellogg did not settle in New York or New Jersey to be by their troubled relatives. They went back to Eleanor's town, Washington, D.C., where Bob worked for the firm of Stone and Webster, a Massachusetts-based engineering and construction company, with which Bob's Wall Street uncle, Henry Ebenezer, had a business involvement. In 1919, Bob left this growing firm for a position as an assistant purchasing agent for the U.S. Housing Corporation. For a time he also worked for a bottled-water producer. The life of adventure seemed well behind Birdseye in 1920, when he switched jobs again and became the assistant to the president of the U.S. Fisheries Association.

This frequent switching of positions seems to have been essentially a search for a new direction. Perhaps he was distracted by the trial. It was not immediately apparent, but the decision in 1920 to work for the fisheries was another one of those serendipitous moves that so often directed Birdseye toward his destiny. The U.S. Fisheries Association was a lobbying group for commercial fishermen that worked on improving the fishing industry on a wide range of matters from proposing legislation to Congress to improving transportation for fresh fish to devising a code with fewer letters so that fishermen could telegraph reports for smaller fees.

This position with the Fisheries Association came as the trial was ending. On March 5, Bob's father and brother were sentenced to prison for two years. Their sentencing was covered in the *New York Times*. Clarence Frank Birdseye Sr. was a powerful, well-respected man and the patriarch of the Birdseye family. His new book denouncing Marxism was about to be published.

On April 30, 1920, Birdseye, aged sixty-five, was received at the Western State Penitentiary in Pittsburgh, prisoner 10849. His occupation was given as lawyer/author. He surrendered his gold cuff links, gold collar buttons, gold penknife, and gold ring and began his two-year sentence. In his mug shot he strikingly resembles the son who was named after him—same balding round head, wire glasses, gray eyes. His eyebrows are a bit thicker, his build heavier, and he had a thick mustache. But in the side-view mug shot the profile is identical.

Though Kellogg was sentenced alongside him, he did not report to the state penitentiary, and it is not clear if he ever served his sentence.

His father's and his brother's convictions must have been devastating to Bob. No one remembers him ever speaking about this event. In fact, the incident is not well known among his descendants. His father served his two years in the Pittsburgh penitentiary and was released in 1922; he died five years later in 1927 at the age of seventy-three—two years too soon to see his namesake son become wealthy and famous. His widow, Ada, moved in with Bob's sister Miriam in Washington, D.C., where Miriam did nutritional studies for the U.S. Department of Agriculture. But by then Bob and Eleanor were gone from Washington.

Those years while his father was in prison, 1920 to 1922, were formative ones for Bob Birdseye. As assistant to the president of the Fisheries Association, he was confronted with the full range of fishery issues, from sea to market. And so he was back to being involved with the two issues that always drove his interest, wildlife and food.

What caught his attention, what excited his imagination, was the problem of getting fresh fish to market in good condition. Most fish lost its value while being transported. "The inefficiency and lack of sanitation in the distribution of whole fresh fish so disgusted me," Birdseye explained twenty years later in a speech at Montreal's McGill University, "that I set out to develop a method which would permit the removal of inedible waste from perishable foods at production points, packaging them in compact and convenient containers, and distributing them to the housewife with their intrinsic freshness intact."

That was the big Birdseye idea. If he could find a way to deliver fish to the customer in the same condition as it landed on the docks, many more people would eat fish, and the fishing business could greatly expand. Birdseye developed a container, inexpensive in its design, that would keep fish chilled until it arrived at market. With Birdseye's box, fish arrived in considerably better condition, but it was still not comparable to fresh fish, and a great deal of it was still lost to spoilage. There had to be a better solution.

Later in life, Birdseye developed a pet theory that the subconscious resembled an electronic calculating machine. "If you feed the right information into it," he would say, "it will quietly go to work in mysterious ways of its own and, by-and-by, produce the answer to your problem."

He kept thinking about his constant struggle for "fresh food" in Labrador. As Birdseye retold the story, after much reflection on the problem following the failure of his container device, "My subconscious suddenly told me that perishable foods could be kept perfectly preserved in the same way I had kept them in Labrador—by quick freezing!"

The more Birdseye thought about this, the more he became convinced this was an idea with a huge potential. In 1922 he left his job with the Fisheries Association and moved to New Jersey, where he persuaded an ice cream company to loan him the use of an area in its plant to conduct experiments in freezing. Eleanor was about to give birth to their third child, Eleanor. Ruth, their second, was a frail child with a cleft palate, a fairly common birth defect, and it was uncertain what medical expenses that condition might require. But Bob and Eleanor, after a quiet five years, were off on another adventure.

To Bob Birdseye, fast freezing was a traditional idea that came from the Inuit. All his life he credited them. But it also has other deep roots in human history. The use of fire and heat developed much faster than the harnessing of ice and cold. This may have been because heat, associated with life, is more appealing than cold, which is associated with death. Or it may simply be that it is much easier to learn how to make fire than to make ice. While human beings were relatively quick to heat their food, it took a long time before they learned to chill it, even though this lack of cold meant that a great deal of food was spoiled and discarded. Spoilage was simply an accepted fact of life. Even in the twentieth century, after commercial freezing was developed, it was a struggle to convince markets that the money saved by less spoilage was a savings to be factored into the cost of freezing. For centuries the primary method to reduce the spoilage in transit caused by warm temperatures was to bring food to market at night. Animals would be slaughtered at the market to avoid having the meat spoil on its way there. Fish were brought to market in ships with tanks to deliver them live.

Ice was largely a luxury for the rich. The members of the Medici court in Florence enjoyed iced drinks. Not knowing how to produce ice artificially, they used natural ice from the mountains in their drinks. They had not advanced one step beyond the Romans, who also served drinks with ice. Pliny invented the ice bucket so that wine could be chilled without being diluted by melting ice. The Romans had icehouses insulated with straw. In China houses for storing winter ice date back to the seventh century B.C.

Frozen food had an inauspicious beginning with Sir Francis Bacon, lord chancellor of England, one of the first English scientists, perhaps the first, and the first martyr to the frozen-food industry. Bacon speculated that a number of "magic" stunts that produced cold were done through the use of salt (sodium chloride) or saltpeter (potassium nitrate) to intensify the cold of snow or ice.

In March 1626, he was riding in a coach through the Highgate area of north London with the physician of King Charles I, and through his window he gazed at the winter snow still on the ground as though the earth were sprinkled in white salt. Though remembered as the great champion of the scientific method, Bacon labored in theory but rarely carried out actual experiments himself. But on this occasion, while speculating with the physician on whether snow could preserve meat the same way that salt did, he suddenly had an urge to try something. Stop the coach!

The two got out and went to a poor woman's house, where they bought a chicken, which the woman killed and cleaned for them. The two scientists went outside, knelt on the ground, stuffed the bird with snow, then encased it in more snow—the same process that Birdseye and his Labrador neighbors used to call snow packing. The cold affected not only the chicken but also Bacon, who became extremely ill. When he was taken to a nearby house, his condition grew worse. He wrote that the experiment in chilling the chicken "succeeded." Only hours after writing the note, Bacon died of pneumonia. Because of this sad incident, it is often claimed that the first frozen food was produced in 1626.

The year after Bacon's death Robert Boyle was born in Ireland to the very wealthy Earl of Cork, who judged him too sickly for school and had him taught by tutors at home. Here is a poster child for homeschooling. Boyle became the first chemist in the modern sense of the word—a researcher who conducted experiments that he meticulously recorded to prove hypotheses about the composition of the physical world. He represented a huge step forward in understanding the concept of elements.

Boyle was interested in understanding everything he could about the phenomenon of coldness. A thorny issue of the day was the question of where cold came from. For a long time many Europeans believed it came from an uncharted island north of the British Isles known as Thule. Aristotle had written that cold came from something called *primum frigidum*. But what was this source? Aristotle said it was water, which Boyle refuted by showing that material that contained no water, such as metals, could be chilled. He also pointed out that the surface of water, the part exposed to the air, was the first to freeze. If water were the source, would not freezing start at the center of the body of water? He also disproved theories that earth, air, and saltpeter were the source of cold. However, he did conduct experiments on how saltpeter and numerous other salts intensified cold. He also challenged Descartes's widely held belief that cold was simply the absence of a drifting substance called heat.

Boyle also investigated the nature of ice, showing that it was water in an expanded form. He conducted numerous experiments to demonstrate that the mass of ice was greater than the volume of water from which it was made.

. . .

America was to become the place where refrigeration and commercial freezing and therefore frozen food were developed. This is striking because Europeans were in the forefront of preserving foods through smoking, salting, and canning. A French chef, Nicolas Appert, invented canning in the Napoleonic era by discovering that food that was heated and sealed in a jar would not rot. The French and later other Europeans had great enthusiasm for canned or jarred foods and even today feature them prominently in deluxe food shops. This may be why they were slow to embrace frozen food, which is still often scorned as "American." But the American embrace of cold as a way to preserve food has also meant that Americans led the way in the manufacturing of refrigerators and freezers.

The first refrigerator was a European invention, but this offers a classic example of the difference between European and American inventors. In 1748, William Cullen, professor of medicine at the University of Glasgow, built a refrigerator based on the known fact that evaporating liquids cause a lowering of temperature. Cullen made a container lose not only its fluid but its air as well, becoming a complete vacuum, with the result that the water in a tank that surrounded the refrigerator froze. He wrote a paper on his work and had it published in a local Scottish journal, but he never attempted to promote the idea.

It took another half century before an American built a refrigerator, but his approach was very different. The device was built by a Maryland engineer, Thomas Moore, who like Birdseye was less interested in scientific theory than in solving the problem of getting fresh food to market. In 1803, he built a metal box for butter, surrounded by ice in a cedar container tightly insulated with rabbit fur. With this box he was able to carry fresh butter from his farm to the Georgetown market, twenty miles away.

Everyone wanted his hard butter rather than the unrefrigerated coagulated grease of other producers and gladly paid his very high prices. Clearly, Moore wanted to start an industry. He patented his box and published a pamphlet about "the newly invented machine called a refrigerator." He invented the word. The industry didn't follow, however, and Moore earned little money from his invention.

The American dominance in freezing and refrigerating started with the fact that America democratized ice. In the eighteenth century most European countries stored ice in icehouses, traded it commercially, and used it chiefly to chill drinks. But it was a luxury used sparingly by aristocrats. In Russia and France ice was controlled by a royal monopoly.

Originally, America started out the same way. The average colonial American had no access to summer ice, but icehouses were a feature of the large Virginia slave plantations. George Washington, Thomas Jefferson, and James Madison all had icehouses for enjoying chilled drinks any time of year. In the North, however, harvesting ice was a low-status and low-wage job; in Philadelphia prison convicts were used. Yet wealthy southerners paid high prices for ice.

Further hampering the ice market was the religious belief that ice out of season was tampering with God's design. There was a similar religious conviction about greenhouses. What must Puritans have thought of Thomas Jefferson, who had both?

But after the revolution, icehouses became popularized,

largely by a man named Frederic Tudor. Tudor was in some ways a Birdseye figure. He was the son of a distinguished lawyer who had clerked for John Adams. The father sent three of his sons to Harvard but not Frederic. In this instance the family had not run out of money, though they would some years later. At the time young Tudor was faced with college, he just thought it was a waste of time.

He started working in a shipping office and, at age seventeen, invented an improved bilge pump. A few years later he started a business shipping ice to the Caribbean, particularly Martinique and Cuba. He shipped it carefully insulated in hay but noticed that Caribbeans, understanding nothing about the nature of ice, let it melt as they walked through the street with it in the midday sun or even dissolved it as they attempted to store it in water. In the 1820s Tudor hired Nathaniel Jarvis Wyeth, another man who had shunned education. Jarvis revolutionized the ice business, first with a saw-toothed ice cutter that made the blocks more regular and produced them more cheaply, and then with the use of sawdust for insulation. Wyeth produced a series of inventions that by the 1830s made Tudor's facility the premier icehouse in America. Tudor's ice, much of it cut from Henry David Thoreau's Walden Pond, was shipped all over the world.

Tudor became a multimillionaire, and his ice became a common international commodity. By the Civil War, Tudor and his competitors were shipping New England ice to fifty ports, in ships leaving Boston every day. New uses of ice were being contemplated. Couldn't it be used to ship meat? The United States had a surplus of meat, and there was great demand for meat in Europe. By the 1830s ice had already transformed the diet of urban Americans. Fresh fruits, vegetables, meat, fish, and milk became increasingly popular. Farmers in upstate New York shipped their milk and other products on ice by train to New York City markets. Fresh fish was brought in on ice from New England.

Americans rapidly became the world's greatest ice consumers, largely because Tudor's and Wyeth's innovations had greatly reduced the cost of ice. Twelve and a half cents could buy a hundred pounds. Southerners became passionate about iced cocktails, the mint julep being one of the most famous. The Port of New Orleans became a major destination for ice, and New Orleans became known for its cocktails. By 1850, New Orleans was buying fifty thousand tons of ice every year, and even though the ice could sell as cheaply as $15 a ton, due to the volume this was important commerce. But nowhere was more ice consumed than in New York City, which used twice as much as New Orleans. Even today in Europe the liberal use of ice in drinks is thought of as an American taste.

By the Civil War ice in America was used not only for drinks but also to store food and keep it fresh. Gradually icebox manufacturers came to understand that the flaw in the Moore model was that there was no circulating air, which helps to cool the ice. This, in fact, is why, as Boyle noted, ice forms first on the surface. After 1845 iceboxes had circulating air and were far more effective.

European scientists continued to develop important ideas, but rarely did they contemplate their commercial application. In 1834, Charles Saint-Ange Thilorier, a chemist in the School of Pharmacy in Paris, was able to apply enough pressure to carbon dioxide to convert it to a solid, carbonic acid. This was an arduous and dangerous undertaking that others had failed at, and Thilorier's assistant had lost both legs in an explosion caused by one of their experiments. But he never attempted to commercialize his discovery, although many years later the same carbonic acid, now called dry ice, became an essential part of refrigeration. It is clear that Thilorier understood the potential of his discovery. He mixed dry ice with snow and ether and produced a temperature of –110 degrees Celsius, or –166 degrees Fahrenheit, which at the time was by far the coldest temperature ever produced artificially. He simply was not interested in practical applications.

Greater progress was made in the second half of the nineteenth century. In 1859, a Frenchman, Ferdinand Carré, produced a machine that made artificial ice by the use of rapidly expanding ammonia, known as a gas-vapor system. Though it took some time for this technology to be accepted by industry, eventually it would mean that ice could be made anywhere. It no longer had to be cut and stored in the North at wintertime. But despite this invention, the ice trade prospered for the rest of the century. Maine became the leading ice-producing state, in its best year, 1890, shipping 3,092,400 tons out of the state.

The United States was not very interested in artificial ice, because American winters—the cold that had once shocked European explorers and settlers—provided ample natural ice. But suddenly at the height of ice production, in 1890, an extremely mild winter—a boon for Maine ice producers but a disaster for those in Massachusetts, New York, and the Midwest—forced Americans to turn to artificially produced ice, which they have used ever since.

Some historians fix the date of the first frozen food as 1875, but there may have been earlier attempts as far back as the Civil War. The first U.S. patent for a frozen-fish process was awarded to Enoch Piper of Camden, Maine, in 1862. He froze salmon by placing it on racks under pans of ice and salt, which lowered the air temperature. After the fish was frozen hard, which took about twenty-four hours, he dipped them in water to give a glassy sheen. More patents followed, and small frozen-fish operations started around coastal New England. Fish was the usual object of frozen-food experimenting, because it was the product that had the greatest losses from spoilage. But it was also the most difficult food to freeze. People could accept a berry that on thawing had gotten a bit soft and juicy, but it is very hard to sell soft and juicy fish.

In 1875, large-scale operations began freezing food in a room insulated with sawdust, the old Wyeth idea, using ice and salt, which is an even older idea. Birdseye used to see this type of operation still in use for frozen bait in Battle Harbour, Labrador, and described the operation in a letter to his parents: "In the upper room cracked natural ice and coarse solar [evaporated sea] salt were mixed and filled into rectangular metallic tubes which passed downwardly through the freezing and storage chamber on the ground floor. Surplus brine was exhausted onto the ground from the lower ends of the tubes." In 1876, American frozen meat was first shipped to England, and by 1881 it was being shipped as far as Australia.

Frozen-food production spread from New England to the Great Lakes region to the Pacific Northwest. By the last decade of the nineteenth century, British Columbia was shipping a million pounds of frozen fish to Europe every year, mostly salmon but also halibut and sturgeon. By 1902 H. A. Baker Sr. was freezing berries in barrels in Puyallup, Washington, and selling them.

But this frozen food had an unenthusiastic public, attracted mostly by low prices, while food critics gave it poor ratings, especially the fish, as did nutritionists. This was an age when prime fish such as salmon and halibut were being landed in enormous quantities and sold at low prices. But only the fish that had not sold while fresh was sold to the freezing companies, so frozen fish started out at a very low quality.

By the 1890s mechanically made ice was increasing frozen-food production. Even before the issue of fast-versus-slow freezing emerged, there was the issue of direct versus indirect— whether or not the food would have direct contact with the refrigerant, which was usually salt in some form. Direct contact was inferior since it exposed the food to salt, and part of the appeal of freezing was as an alternative to salting food. In the beginning of the twentieth century there were many experiments with different chemical formulas and equipment for freezing. The first industrial freezer in the United States was the Ottoson brine freezer. The food was submerged in a brine solution at the freezing point. The brine had the same percentage of salt as the cellular composition of the food, which was supposed to keep the salt from being absorbed by the food. But this equilibrium was nearly impossible to maintain, and often salt was absorbed into the food, ruining the product. In 1921, Paul W. Peterson built the first "indirect freezer"—a machine in which the refrigerant and the food never come into contact. The food was packed in containers and immersed in a liquid refrigerant. In 1923, the year Birdseye started his first frozen-food company, Gordon F. Taylor developed a new kind of direct-contact freezer in which whole fish moved on a conveyor belt under a spray of cold water and then were frozen with brine. The frozen fish were then sprayed with water again to wash off the brine and give them a crystalline glaze of ice. From Birdseye's point of view, the interesting idea in Taylor's machine was the use of a conveyor belt for mass production.

By then the advantage of fast freezing was well known in science, and even industry had experimented with it. The earliest commercial fast-frozen food, which was whole fish in salt and ice, had been produced in 1915 while Birdseye was in Labrador.

By the 1920s European scientists had expounded on fast freezing and the principles of crystallization, but in reality most frozen food was still slow-frozen and of very poor quality. In fact, as more frozen food became available, the quality got worse. The problem was that because of its bad reputation, it only fetched the lowest prices and so had to be made from the cheapest, poorest-quality fresh food. New York State even banned the serving of frozen food in its prisons. State laws were passed to try to protect consumers from the terrible frozen food. In the state of New York slow-frozen food could not be sold unless a store posted a sign over its entrance in letters a minimum of eight inches tall stating, "Frozen Food Sold Here."

Perhaps the reason Birdseye didn't arrive at the idea of freezing sooner was that frozen food was so thoroughly associated with poor-quality food. Frozen fish was presented at the London 1883 International Fisheries Exhibition, a highly influential gathering of all the North Atlantic fishing nations that shaped attitudes about everything from eating fish to managing fisheries. The fish received horrible reviews, the press describing its smell as a "stink" and its look as "withered." Forty years later this was still the image of frozen food.

There was a minor breakthrough in acceptance of frozen food during World War I when the U.S. Army bought relatively small quantities of frozen fish and chicken to feed soldiers in

Camp Venustus in the Bitterroot in 1910. *Left to right*: Willard V. King, Paul Stanton, Clarence Birdseye in three-piece suit, and Robert Cooley.

Birdseye looking happy in fur in Labrador, circa 1912.

Wilfred Grenfell with his team of dogsleds, 1912.

Grenfell, satirizing Birdseye's hunting prowess in a 1912 sketch that Birdseye pasted into his diary for October 10, 1912.

Grenfell's drawing of the crew of the *Strathcona* on bended knee, giving thanks to the hunter Birdseye who brought them geese. This drawing, too, was pasted into Clarence Birdseye's journal entry for October 10, 1912.

Letterhead torn off of hotel stationery from the Taconic Inn—where Birdseye and Eleanor spent their honeymoon—pasted into his diary without comment.

Clarence and Eleanor's first home, which they had built in Muddy Bay, Labrador.

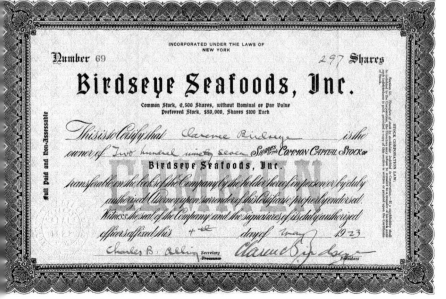

Stock certificate in Clarence Birdseye's name for his original company in New York, Birdseye Seafoods, May 1923.

Birdseye with Eleanor and Henry, circa 1929.

Marjorie Merriweather Post, whose General Foods bought Birdseye in 1929.

This man has wrought a miracle!

Clarence Birdseye
eminent scientist and explorer, and the inventor of the Birdseye Quick-Freezing Process

Portion of ad for Birdseye Frosted Foods that ran in the March 2, 1930, issue of the *Springfield Republican*.

The Birdseye building in 1930, which has become a landmark in Gloucester Harbor.

Birdseye riding one of his horses in West Gloucester, as seen in one of his home movies from the 1930s.

Birdseye chasing finback whales off of Gloucester with the tagging harpoon he invented, also from a 1930s home movie.

October 7, 1944. Birdseye is feeding carrots into a machine that trims them to the appropriate size for dehydration.

Dehydrating carrots in the summer of 1943 in his laboratory on the second floor of the Birdseye building in Gloucester Harbor. *Left to right*: A. Pothier, Clarence Birdseye, and Helen Josephson Schuster.

Clarence Birdseye in his office in 1943.

The dining room of the Birdseye home on Eastern Point, Gloucester, in 1942 or '43. Clarence Birdseye has his back to the camera, Eleanor is opposite him facing the camera, his daughter Ruth is to his left, and his son Henry is to his right.

Birdseye loved serving lobster to guests in his Eastern Point home and called it a "lobster feed." This one was in 1947.

The Birdseyes in their home in Peru, winter 1954.

Birdseye feeding ocelots in Peru.

Birdseye in 1955 or '56, after he returned from Peru.

the United States. This led to a considerable amount of talk about frozen food after the war, though few people actually ate it. By the time Birdseye began his experiments with freezing food in 1923, frozen food had been around for fifty years, and all the technology he needed for the kind of production he envisioned was available. But, by his own confession, he knew almost nothing of this research and development. He did not know Ottoson's machine, or Peterson's, or most of the studies and debates. He only knew what he had learned about freezing in Labrador.

In 1923 he organized his own frozen-fish company in New York by selling $20,000 in stock. Birdseye looked everywhere for investors for his big idea. At the time Henry Ford had expressed interest in frozen food, and Birdseye made an effort to get his backing but got no response. However, the Clothel Refrigerating Company, which produced refrigerators for the navy, was interested in the new company, called Birdseye Seafoods, and provided space on White Street a short distance from the Fulton Street Market, where wholesale fresh fish was bought and sold.

The originality of Birdseye's work was as much in the packaging as the preparation of food. He was still primarily interested in the issue of how to get the product from wholesaler to retailer to consumer in a sanitary, efficient, convenient package. He cleaned fish, chilled but did not freeze them, and shipped them in well-insulated fiberboard boxes. If his boxes had worked better, he might not have gone into freezing. But he quickly learned that this packaging was not adequate to ensure a fresh product. That was when he started thinking about the food of his Labrador years.

Perhaps the most famous Birdseye legend is how he went to

a store and spent $7 on salt, ice, and an electric fan and with these reproduced the Labrador winter and froze a fish. There is an element of truth to this. This is indeed how Birdseye worked. He found a few banal household items and solved a problem. And he loved wonder-of-science-type demonstrations. It was one of the things that would later make him popular with neighborhood children. He would dehydrate a carrot before your very eyes with household equipment like an upside-down electric coffeemaker. So he probably did perform this demonstration, though where and for whom is not known. But had this been all he accomplished with freezing, we would never have heard of him. It was a great trick, but just how to freeze, even fast freeze food, was already known. He was not just out to freeze; he was out to create an industry, to find a commercially viable way of producing large quantities of fast-frozen fish.

The problem with the existing frozen food was not only that it didn't freeze quickly enough but also that it wasn't cold enough. Birdseye estimated that a piece of meat was not truly frozen until it reached −96 degrees Fahrenheit. Existing commercial freezing was done at about 25 degrees, only a few degrees below the freezing point.

By this time there had been a great deal of experimentation with different formulas for arriving at freezing temperature. Pressurizing gases into solids as Thilorier had done with carbon dioxide continued to yield extraordinary subzero temperatures. In 1877, Louis-Paul Cailletet liquefied oxygen and nitrogen. In 1898, only twenty-five years before Birdseye set up his company, James Dewar, a Scot, the inventor of the thermos bottle, liquefied hydrogen and reached a temperature of −250 degrees Celsius, or −418 degrees Fahrenheit. This was the age to be rethinking freezing.

Most people working in freezing did not get involved in science as sophisticated as this. They experimented with different salts and configurations of ether. Clothel's navy refrigerator used ethyl chloride. Others used potassium nitrate or saltpeter, as Bacon had suggested centuries earlier, or calcium chloride.

Birdseye always chose low technology over high, and so he too investigated salts. For all their low technology salts are a baffling world. Anyone wishing to understand the struggle engaged in by Bacon and Boyle to move thinking away from alchemy—the belief that matter changes by magic—to science—the belief that it acts in accordance with provable natural laws—should look at salts. Though salt acts by natural laws, it can do so many different and seemingly contradictory things that it appears to operate by magic.

There are many different salts. Sodium chloride, which is the salt we eat, potassium nitrate, potassium chloride, calcium chloride, magnesium sulfide, and even monosodium glutamate are all examples of salts. A salt by definition is a compound created by the neutralization of an acid and a base. Both are extremely unstable—one because it lacks an electron and the other because it has an extra one. So they are, like a good marriage, attracted to each other because they solve each other's problem, complete each other, and result in an extremely stable compound.

But this stable compound, the salt, has many unusual properties. Not only does it preserve food, make water buoyant, put out fires, and keep cells functioning, but it can perform a number of seemingly contradictory tasks relevant to the story of freezing. It makes water boil at a higher temperature but also makes it freeze at a lower temperature. It does these two seemingly opposite things for the same reason, because the salt mixes with water and carries properties that react differently to heat and cold. Salt has a higher boiling point and lower freezing point than water. Salt melts ice—implying heat—but it also has just the opposite effect; while it is melting ice, it absorbs heat, making everything around it colder. The ice melts because the salt is establishing a lower freezing point. But as the heat is absorbed, it chills the area to that new lower freezing point. How low a temperature it descends to depends on how much salt and what type of salt is used.

Three hundred years before Birdseye started his freezing company, Bacon had written about this, though he did not entirely understand it. It became the standard way of freezing, and it did not take long for Birdseye to learn how to drop the temperature. He experimented with different salts and different solutions and arrived at different temperatures. Much of the experimentation had gone on in his home in Yorktown Heights, in Westchester County, New York, before he acquired commercial space. He was initially interested in dry ice. Years later Birdseye would recall, "Production of perishable foods, dressed at the point of production and quick frozen in consumer packages, was initiated, so far as I am aware, in the kitchen of my own home late in 1923 when I experimentally packaged rabbit meat and fish fillets in candy boxes and froze the packages with dry ice."

He eventually settled on calcium chloride in a solution that quickly dropped the temperature to −45 degrees Fahrenheit, which was the temperature of some of the coldest days in Labrador. He discovered that all that was necessary for fast freezing is that the temperature pass very quickly out of the first freezing temperature range, which is from 33 degrees to 23 degrees Fahrenheit. Below that, crystallization is rapid, and crystals become very small. To speed up freezing, smaller amounts had to be frozen. Before Birdseye freezing was an entirely wholesale business, and huge blocks of food, including whole sides of beef, were frozen. Fruit was frozen into four-hundred-pound blocks. The food was placed in a barely freezing environment and frozen over a number of days. Birdseye, like Gordon Taylor, froze one fish at a time and froze it quickly. His quick-frozen fish went on the market in 1924.

His packing and freezing machine weighed twenty tons, considerably larger and heavier than Clothel's two-ton refrigerator. In 1924, Birdseye was awarded his first patent for this fish-freezing process. In his patent application dated April 18, 1924, he made the following claim: "It is well known that fish like many other dietary articles has been frozen for the purpose of suspending, or preventing, decomposition through oxidation, bacteriological or other action, but the present invention goes much further than that and accomplishes results heretofore unknown in the seafood industry."

If Birdseye didn't pioneer actual freezing, he had to pioneer most everything else in his process. He experimented with the different heat-insulating properties of fiberboard, cardboard, cork, and wood. His ideas were so new that the National Bureau of Standards could not furnish him with any information on the insulating properties of fiberboard. No one had tested it, and so he did it himself. His first patent makes clear that his concern was as much with packaging as with freezing itself. The patent goes into elaborate diagrams and explanation of the packaging, insulated fiberboard boxes, and was the first of many patents Birdseye was awarded for packaging.

One of Birdseye's first true innovations was packing fish in corrugated cardboard, fiberboard, and eventually waxed cardboard, rather than the traditional balsa-wood boxes used for salt fish. Balsa wood worked well enough, but he needed to produce frozen food inexpensively, and the wooden boxes cost $1 for every pound they held. They were supposed to be returnable, but customers seldom bothered. Later in 1924 he patented his second, improved box. He was very concerned with eliminating all air pockets within the package of fish because bacteria, which led to decomposition, could grow in these spaces. In his original 1924 process he called for the packing of fish fillets, which was unusual at the time because the fish had to be filleted and skinned by hand.

By early 1924 he was out of money. But he believed he was on the path to an important innovation. He had solved a few of the practical problems, but he needed more time and more money to solve the others. Though they had three small children to raise, Bob and Eleanor knew what to do. They sold their life insurance policy for $2,500 or, according to some, $2,250. This may seem like a foolhardy thing to do, selling off all you have and sinking it into an experimental industry, but Bob always said that the only way to succeed in life was to be willing to take chances. During their long marriage, Eleanor never doubted Bob. She used to explain to their children, "Dad was born ahead of his time. There is so much going on in his head the world can't even catch up." Their daughter-in-law Gypsy, Kellogg's wife, said of her, "I call her Saint Eleanor. He was just the most fortunate man to have had her for a wife."

Birdseye Seafoods was broke within a year. New Yorkers just didn't want to eat frozen food. But now Bob Birdseye had a different idea. He and Eleanor sold their house in Westchester and with their three children and a fourth expected moved to Gloucester.

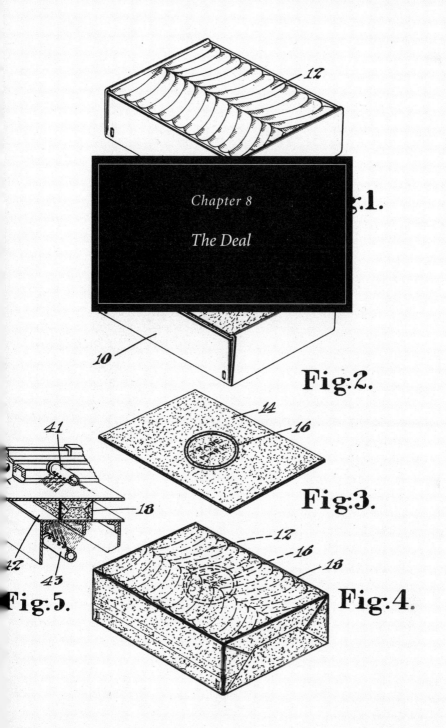

12

Chapter 8

The Deal

Fig.1.

10

Fig.2.

14
16

Fig.3.

41

18

12
16
18

42

43

Fig.5.

Fig.4.

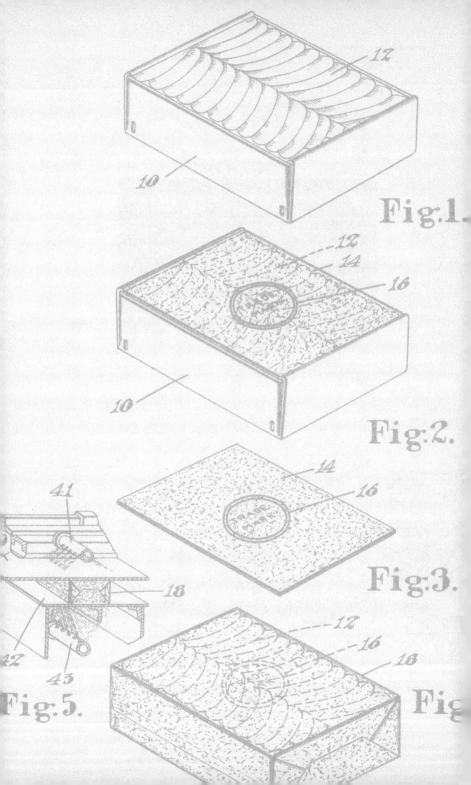

Fig.1.

Fig.2.

Fig.3.

Fig.5.

Fig.

When the Birdseyes arrived in Gloucester, the city had just celebrated its three hundredth birthday. Founded in 1623 as an English fishing station, it remained, and still is today, one of the leading fishing ports of the United States. Its deep and extensive harbor, on the headlands of a peninsula but in the sheltered lee of the wind, makes it the ideal port for New England's richest fishing grounds.

By 1873 the population had grown to 15,397, making it the most populous town in Massachusetts, and so a new charter designated that it was now a city. When the Birdseyes arrived in the mid-1920s, it was only slightly larger, and most of the population, directly or indirectly, lived off the fisheries. In addition to the fishermen themselves and their families, there were the lumpers—the dockworkers. Kids could be lumpers if they were big and strong enough and by working very long hours could earn enough money, if they were frugal, to buy a house when they got married. There were also the ancillary trades that repaired ships, made iron fittings, and manufactured oilskins for fishermen to wear. Glue was made from fish skin, later sold nationally by William LePage.

It was a working town with a tough waterfront of bars and merchants, and the harbor was crammed with the masts and canvas sails of some 150 working fishing schooners. There were also fat-hulled, square-rigged barks from Sicily that arrived with Trapani salt. The Gloucester fish industry used enormous quantities of salt—one of the things Birdseye was about to change.

The schooner, a fast-sailing vessel, sleek and rigged not with cross spars but from bow to stern so that it would be swift and maneuverable and sail close to the wind, was an eighteenth-century Gloucester invention designed for fast voyages to the Georges Bank fishing ground and a fast sail home again. Schooners were so swift and so beautiful that they were redesigned as racing yachts. The America's Cup race was originally a contest between Gloucester and Nova Scotia fishermen.

Gloucestermen were so in love with their wooden-hulled fishing schooners, most of them built in the nearby marshes of Essex, that in the 1920s, well into the age of engine-powered steel-hulled fishing boats, the Gloucester fleet was still mostly schooners, and fishermen continued to use them into the 1950s. But they were built for speed and not safety. They flew large topsails high on the masts and would easily blow over in a storm. They fished bottom fish, mostly cod and halibut, with hand-hauled longlines with baited hooks from small two-man rowboats called dories. The dories could capsize on a high sea, or sometimes just catch too many fish and sink to the bottom from the weight, or sometimes get lost in a fog and never find their way back to the mother schooner. Some years hundreds of men were lost at sea.

So they were a tough people, living a hard and tradition-bound life, and accustomed to tragedies that bonded them into a closely knit, hardworking blue-collar community of skilled workers, and for Birdseye this made Gloucester an ideal place for starting a new industry.

But Gloucester also had a beauty, a quality of sunlight muted through ocean haze and reflected on the water that had always attracted great painters. Marsden Hartley, Edward Hopper, John Sloan, and Stuart Davis were all working in Gloucester, bemusing the locals with their easels and palettes, when the Birdseyes arrived.

There was already an inventor in town, John Hays Hammond Jr., the brother of Birdseye's former backer in the fur trade Harris Hammond. Although John Hays Hammond was less than two years younger than Birdseye, and even though both were in Gloucester, they were worlds apart. Not surprisingly, it was much easier for Hammond to get Hammond money behind him than it was for Birdseye. When Hammond wanted to start a company in Gloucester, his father gave him $250,000. Edison and Bell had been right when they saw potential in young Hammond. He invented remote control. In 1914 he sent an unmanned yacht guided by radio signals from Gloucester Harbor to Boston and back, a straight 120-mile course. It disconcerted fishermen encountering the ghost ship at sea. During World War I he invented a system whereby a remote-controlled ship could home in on an enemy's searchlights. He also invented a system to protect the remote control from enemy interference and then worked on the first radio-guided torpedo. He also invented an eyewash, an altitude-measuring system for airplanes, a magnetic bottle cap remover, a meat baster, and a "panless" aluminum stove on whose surfaces food was cooked. In one respect he was much like Birdseye, constantly coming up with ideas for whatever problems confronted him.

There is no record if Birdseye took an interest in the panless stove. He might have been interested in Hammond's cure for baldness, but it didn't work. By the end of his life Hammond's more than four hundred patents even outdid Birdseye. Most of Hammond's important inventions were high-technology electronics. Hammond was also far wealthier and lived far more grandly. At the time Bob and Eleanor were settling in Gloucester, Hammond was building a castle on the west side of the harbor. He said he built the castle because it reminded him of some of the estates he had lived on as a child in England. His castle was to have a drawbridge and a ten-thousand-pipe organ.

The Hammonds were not the only wealthy family in Gloucester. The city is on the tip of Cape Ann, the last stop on what is to Bostonians "the North Shore." Since the nineteenth century, affluent Bostonians had been vacationing or building homes along the North Shore. But many preferred to go no farther than the neighboring town of Manchester-by-the-Sea. They called the next stop "Gloucester-by-the-Smell." Gloucester was built on a series of inlets, coves, and peninsulas, but to find the center of town, one had only to sniff. Lining most of Gloucester Harbor, which includes downtown Gloucester, were so-called fish flakes, many of them owned by the Gorton-Pew Fisheries, a merger of the two largest seafood companies in Gloucester. Flakes were rough-hewn wooden racks upon which splayed and salted cod was laid to dry. Salt cod was Gloucester's biggest product, and it had to spend weeks drying in the sun, giving off a rich fishy odor. For people who grew up in Gloucester, it was just the smell of home, but it made a strong and often negative impression on visitors.

Birdseye moved to Gloucester to build a frozen-seafood company in a place where very fresh fish was readily available. Although he intended to sell frozen fish, the real goal of his company was to develop and patent ideas for frozen food.

His company was to develop machinery and processes, patent them, and license them to other food companies. In other words, he was not as interested in founding a company as in launching a whole new industry.

The exact date when the Birdseyes moved to Gloucester is uncertain, but all the evidence points to 1925. Some accounts, including in the local newspaper, the *Gloucester Daily Times*, state 1923, which does not seem possible, because he was clearly running the New York company that year. *The New Yorker* in a 1946 article had him moving there in 1924. Others have said 1925. But the birth of his last child, Henry, in April 1925 does not appear in Gloucester birth records. His 1924 and 1925 patents list his residence as Yorktown Heights in Westchester, and only in 1926 is he listed as being from Gloucester. But Gloucester city records show that by the summer of 1925, probably starting in May or June, the Birdseyes were living in a comfortable house with a wide porch, 1 Beach Road, near an exclusive golf club. This was not a particularly deluxe neighborhood, but the neighborhoods away from the harbor were considered more desirable because they were away from the smell. The house was a spacious middle-class home in a town where people were still living in shacks by the waterfront. The Birdseye home was an easy walk to what is known as the Back Shore, the part of Gloucester on the open Atlantic, with rough and majestic granite boulders and long sand beaches.

A letter that Bob wrote to one of his backers, Isaac Rice, shows him as a warm and generous family man enjoying a Gloucester summer. He wrote to Rice:

> This would be a wonderful place for you and the family to spend at least a short vacation. There is a won-

derful beach for both grown-ups and children within
about five minutes walk of our house and our two
older kids spend about six hours a day in their bathing
suits. Remember that our house is open to you at any
and all times and that we will be tickled pink to have
you come with as many of your family as you like for
as long as you like.

Bob clearly loved the historic industrial fishing port and
was to spend more years there than any other place in his life.

With his usual charm and infectious enthusiasm he brought
into his company Wetmore Hodges, a young Wall Street
investor. Hodges brought in two colleagues from J. P. Morgan, Isaac Rice and Basset Jones. Together the three invested
$375,000, which had the spending power of $4.6 million in
2010. Two additional investors, William Gamage and J. J.
Barry, had financial connections that increased the company's chances of raising capital. Though Hodges was a young
man, bringing him in was an important step. Hodges knew
many powerful New York financial people, and his father was
Charles Hodges, a vice president of the American Radiator
Company. Bob had also interested the father and the American Radiator Company in the future manufacturing of freezer
units. Birdseye was beginning to lay the industry's foundations. His new company, the General Seafoods Corporation,
was small but with numerous investors and backers interested
in gambling on his new ideas. It was going to take a considerable amount of money and business contacts to develop this
little company into a whole new industry.

Bob's brother Kellogg, who served as a company vice president, moved to Gloucester with his wife, May, who was from

Massachusetts, and their two children. They lived in a historic district in the center of town. But he was only there in the early years of the company, from 1927 until it was sold in 1929. Then Kellogg and his family relocated to Boston. The Birdseye family no longer had a New York–New Jersey nucleus. Roger was in Flagstaff, Arizona; Henry was a New Yorker but had a business in Detroit; while Miriam and Ada were in Washington, D.C.

In 1926, Birdseye's company bought a building on a peninsula in the downtown harbor front known as the Fort, which was one of the oldest neighborhoods in Gloucester. During the American Revolution and the War of 1812 there was an actual fort there guarding the harbor from the British. It was a hill of winding waterfront streets. On Commercial Street by the harbor docks most of the buildings were used for fish processing. This building was unusual for the Gloucester waterfront because it had an elevator, which was housed in a tower. The elevator could carry fish to flakes on the roof. But Birdseye used freezers, not fish flakes. The Birdseye tower has become iconic in the Gloucester Harbor view. The first floor was used to develop frozen-food experiments, and the second floor had offices for developing and planning their future.

Bob's ambitions for the future were not small. He named his company General Seafoods Corporation because he imagined frozen food, still a small oddity, becoming a huge industry in which his company held a place comparable to General Electric and General Motors in their fields. Once the industry was well established, his company, according to Birdseye's plan, would be renamed General Foods. His first manager, Joe Guinane, recalled in 1960 that Birdseye was tirelessly experimenting: "There was not a frozen food item that I do not recall Birdseye experimenting with at one time or another in the early days. He tried all kinds of things—not only seafood, meats and vegetables, but baked and unbaked goods as well as cooked foods." He envisioned the entire future of frozen food.

General Seafoods began with haddock fillets frozen in rectangular cardboard boxes. It produced thousands of pounds of the boxed frozen fillets in the plant at the Fort. Bob's omnivorous tendencies were once again in full swing. He prowled the docks for anything unusual that he could try freezing. He let the Gloucester dragger captains know he was interested. He froze whale, shark, porpoise, and according to one account he even found an alligator to freeze.

Guinane was an engineer who had been working with J. J. Barry, an investor already thoroughly seduced by the irrepressible Birdseye. Barry talked the engineer into joining the venture. "I felt I was taking a rather daring step," Guinane later recalled. But who could resist Bob Birdseye? Guinane said, "I think the primary qualities that made Bob Birdseye what he was, were the combination of a superb restless mind, an insatiable curiosity, enormous persistence, and the total lack of mental blocks."

Though it did pack a great deal of frozen fish and tried to sell it, the company was more involved with developing machinery. That was why Birdseye had chosen Guinane, an engineer, to manage his seafood company.

In addition to freezing machinery and packaging, they worked on many other devices. Barry developed an electronic machine for trimming fish. Birdseye invented a new machine for the old technique of brining, lightly exposing the fish to salt water to accelerate the freezing process. Birdseye also invented a new fish-scaling device. Other electric fish scalers existed, but they would stall out when hitting a fin, while the Birdseye scaler didn't. They built a system that picked up fillets, washed them, and brined them. Birdseye believed he could improve the quality of products if he could develop a controlled mechanized system.

In July 1925, not long after he had set up his Gloucester company, he wrote to Rice that his filleting machine filleted twenty-three and a half haddock a minute and he was anticipating a new and better motor that might do as many as fifty fish a minute. Filleting machines were the new big thing of the 1920s seafood business, and by the 1930s fillets, fresh and frozen, would dominate the seafood market. Representatives of filleting operations at the Boston Fish Pier and from Nova Scotia came to Gloucester to look at Birdseye's work. The men from Boston told him that in the next twelve months the Boston Fish Pier planned to produce and sell 10 million pounds of haddock fillets.

Originally, General Seafoods' freezing process resembled what Birdseye had been doing in New York. Fish fillets were packed tightly in five-by-three-by-two-inch cartons, his first patented process, and then a number of cartons were placed in a long metal holder that was immersed in freezing calcium chloride.

Of the more than two hundred Birdseye inventions, perhaps the most important was the one he applied to patent on June 18, 1927. It was patent 1,773,079, and it truly began the frozen-food industry. Birdseye begins the application by asserting:

My invention relates to methods of treating food products by refrigerating the same, preferably by "quick" freezing the product into a frozen block, in which the pristine qualities and flavors of the product are retained for a substantial period after the block has been thawed.

His invention describes every step of the process, including how to pack fish fillets in a box. Among the advantages he claimed for his method was that it was "indirect"—the product has no contact with the refrigerant—and that the frozen food came out in a package suitable for marketing to the consumer. He claimed to have invented for the first time ever a way of producing frozen food "as a practical, commercial article of commerce."

He accomplished this by packing food very tightly to eliminate air, in cartons holding a two-inch-thick block, and pressing these cartons between two plates that were kept between −20 and −50 degrees Fahrenheit. The carton was held between the freezing plates for seventy-five minutes. The blocks not only froze the food solid but compacted it into a tight regular rectangle, assuring complete freezing in a short period of time and no spaces for bacteria to enter.

Without having made a single scientific breakthrough, Birdseye had developed a process for freezing food on which an entire new industry was founded. It was called multiplate freezing. It remained the basic commercial freezing system for decades. In 1946, *The New Yorker* claimed that the only change in freezing since Birdseye's invention was that the plates were now hollow. But in reality, hollowing the plates and putting an ammonia-based refrigerant inside were only two of numerous Birdseye improvements that came after the initial invention.

He had solved most of the problems that he had pondered. Multiplate freezing could produce a large amount of frozen food, cold enough and frozen quickly enough; the freezing was indirect, the air-space problem was eliminated, and the final boxed product was in a convenient form suitable for marketing. But clearing away these problems only clarified how many other obstacles stood in the way of General Seafoods becoming the General Motors of food. It was not an unprecedented problem in the world of inventing—inventors who were too far ahead of their time, producing cars without paved roads, telephones without telephone wiring. The classic example was Thomas Edison, who developed a lightbulb that would have been commercially viable if only there was an electric grid to hook it up to.

Still ahead of Birdseye was the Adam Trask problem. Trask was a fictitious character in John Steinbeck's 1952 novel, *East of Eden*. In the novel Trask learns in the early twentieth century of a mastodon that has been preserved for millennia frozen in ice and concludes that frozen food is the future. He studies everything he can find on freezing and bacteria and manages to freeze a trainload of lettuce and ship it east from California's Salinas valley. But a snowslide delays the train over the Sierras, the cars cannot protect the lettuce against unseasonably warm weather in the Midwest, and there is a further delay in Chicago. There is no infrastructure for shipping frozen food, and the entire shipment is ruined, Trask loses a fortune, and everyone in Salinas laughs at him while he stubbornly insists that frozen food is the future.

This was now Birdseye's dilemma as well. Because Birdseye had been more focused on developing equipment and industrial processes, he had not found solutions to the problems that would be encountered in distributing frozen food. But that first summer of 1927 he froze 1,666,033 pounds of seafood and was faced with the problem of what to do with it. His company wanted to ship it to the Midwest, where consumers had no fresh seafood. But there were no trucks or train cars for frozen food, and the few warehouses for frozen food to ship to were not cold enough. There was no distribution network capable of dealing with the problems of frozen food, and retail stores had no equipment cold enough for holding or displaying top-quality quick-frozen food.

Then too there was the daunting problem of getting the consumer to accept the product. Just as had happened to Adam Trask, people laughed at the idea of frozen food. Birdseye's food was completely different from what people knew as frozen food at the time, which might have been an advantage considering the low reputation of that food, but most people found the new ideas strange and implausible. To most people, if meat was frozen, it was a whole side of the animal or whole chickens. Frozen vegetables were an unheard-of idea, and the only frozen fruits most people had ever heard of were large barrels of frozen strawberries, most of which were purchased industrially for making strawberry ice cream or jam and jelly. Fish in fillets was also still a new idea. Birdseye's company placed white notices inside the top and bottom of each box of fish that stated in red lettering, "The product in this container is frozen hard as marble by a marvelous new process which seals in every bit of just-from-the-ocean flavor."

Birdseye had also drifted into the middle of a central

debate of his time. Today there is the issue of whether electronics are destroying industrial jobs. But in the 1920s the issue was whether industry was destroying the livelihoods of artisans and craftsmen. Labor unions that represented butchers, seafood processors, and poultry workers came out early and emphatically against frozen food. They believed that if this new idea caught on, food would all be processed in frozen-food factories, and their workers would be put out of business. Birdseye always denied this, but of course they were right, although frozen food has not hurt them nearly as much as supermarket chains. The canning industry also feared frozen food, and once Birdseye established the idea that frozen food would come in rectangular cartons rather than a container that can makers might produce, they also turned against him.

New ideas generate a great deal of distrust. Railroads were unenthusiastic about getting into the transport of frozen food because they imagined huge liability for shipments that accidentally thawed. Public health officials, and even some scientists at the U.S. Department of Agriculture, where Birdseye liked to think he had friends, opposed frozen foods. If it was true that fast freezing preserved everything in its natural state, which it would return to upon thawing, then, they all reasoned, pests and diseases would be preserved in the freezing process and shipped around the country in frozen food.

Birdseye bitterly complained that every time he launched a new product, he was forced to spend a great deal of badly needed research and development money at the Massachusetts Institute of Technology proving it was safe to eat. He proved that freezing killed trichinae, which causes trichinosis, before he was allowed to sell frozen pork and that it killed corn borers before he could get approval for frozen corn. Vegetables were held in particular suspicion by health authorities because, unlike the heating process in canning, freezing methods offered no sterilization.

Almost everything in the packaging process had to be invented. There was not even a strong, transparent, waterproof wrapping material available. This was more than ten years before plastic was invented. The company started with wax paper, but it gave little protection from moisture and completely deteriorated during thawing. Then it used a vegetable parchment, which did stand up to the thawing process but then stuck to the box and also was not at all waterproof. The company coated the parchment with paraffin, and it no longer stuck with thawing, but it dried out in the freezer. It bought cellophane from France. This was a fairly new product that had generated some excitement. The Swiss inventor Jacques E. Brandenberger made it from chemically treated cellulose and had his major breakthrough in 1912, when he discovered that adding glycerin made it pliable. But Birdseye discovered that it disintegrated when it came into contact with wet fish. He persuaded the DuPont Company to make cellophane with a waterproof coating. The chemical company saw a commercial potential to this product and went into production, even though at the start General Seafoods of Gloucester was its only customer. Then cigarette companies bought it, and cigars started to come wrapped in it, and soon cellophane wrappers were a standard feature of American consumer goods—another little-known Birdseye influence on our world.

Even new kinds of ink and glue had to be developed because

frozen-food cartons became wet as soon as they began thawing, and so everything had to be waterproof.

For all this innovation General Seafoods still had no adequate transport or warehouses, and even if it did manage to get its frozen food to retail stores, these stores had no way of keeping the food frozen. Sometimes General Seafoods even sold food directly to a large family. Stores that bought the frozen food often thawed it out and sold it as fresh food.

In 1928, the first retail store freezer became available. According to Birdseye, it cost as much as the entire rest of the average grocery store.

Birdseye developed an improved method of freezing—the continuous belt freezer. Instead of cooling plates, this machine had metal belts chilled with calcium chloride spray at −45 degrees Fahrenheit. The belts could be adjusted to press the packages tightly between them. The two-inch-thick belts froze the package from the bottom and the top so that it froze in very little time.

This was intended not only for greater production but for a future in which all kinds of foods were frozen in different-sized boxes. A space between the belts was adjustable so that the packages would always be tight-fitting. It was at first difficult to find the right belts. Birdseye started with Swedish belts made of watch spring steel, which did not hold up well to brine. He then tried copper, then bronze, then Monel metal, a corrosion-resistant silver alloy. Finally, the solution was stainless steel, a steel alloy with chromium to resist corrosion. Stainless steel had at this time been in development for more than a decade, but it was not widely available until the 1930s. When Birdseye first explained the machine to Isaac Rice, Rice called it a Rube Goldberg contraption. But it was a contraption with a future, an improvement on multiplate freezing—a machine good enough to be the basis of a new industry.

One of the drawbacks was that the Birdseye machine was fifty yards long and weighed twenty tons, and it could only be operated in a cold, insulated room. The company bought a neighboring building in 1928 to house the huge machine.

None of these new ideas out at the Fort in Gloucester were happening in a vacuum. When the Birdseye group struggled to invent a new food industry, it was not as wild a dream as it might seem today, because it was at the dawn of an era when industry was reinventing food and putting new tastes on the market packaged in new ways, seemingly every day. In 1928, Kraft put Velveeta cheese on the market. James Lewis Kraft, born near Fort Erie, Ontario, in 1874, was a Chicago cheese merchant with a set of problems that would become familiar to Birdseye. Cheese was sold under unrefrigerated glass domes, with slices cut off the wheel or block. Heat deteriorated the cheese, and air dried it, forcing retailers to cut the dried end off every time they served a new customer, wasting a large part of the block. If retail stores had had better refrigeration and a good airtight wrapper like plastic wrap had been available, processed cheese might never have been invented. Processed cheese was originally a Swiss invention by Walter Gerber in 1911. Additional salts, such as sodium phosphate, potassium phosphate, tartrate, and citrate, make the cheese more resistant to spoilage by keeping the fats and solids from separating. Kraft began experimenting with a cheese-based factory blend. He made a fortune selling canned processed cheese to the army during World War I and continued developing the product and the company after the war.

In part he combated spoilage by packaging small amounts of cheese. Velveeta, wrapped in tinfoil and packed in individual wooden boxes, was his crowning glory. It melted so smoothly it seemed like velvet. Kraft originally was not trying to make cheese melt well, but the meltiness of his processed cheese—because it would not separate with heat the way real cheese does—changed the way Americans ate. It made melted cheese ubiquitous in the American diet, including macaroni and cheese, cheeseburgers, grilled-cheese sandwiches, and au gratin everything.

The same year as Velveeta came out, Walter E. Diemer of the Fleer Chewing Gum Company invented bubble gum, to many kids the greatest thing since sliced bread, which itself was invented that same year when Otto Frederick Rohwedder's commercial slicing machine was first used in Chillicothe, Missouri. The slicer debuted on July 7, 1928, according to a story that day in the *Chillicothe Constitution-Tribune*, which covered the launching. In 1929, the year Harland Sanders opened his first fried chicken restaurant in a gas station in Corbin, Kentucky, the soft drink 7UP was first marketed as Bib-Label Lithiated Lemon-Lime Soda.

Nor was Birdseye the only one experimenting with packaging. In addition to Kraft's wrapped cheese, in 1929 milk was first sold in cartons instead of bottles, and the British scientist E. A. Murphy whipped latex rubber in a kitchen mixer and created foam rubber.

By 1928 most of General Seafoods' impressive production of frozen food had not been sold. The company was learning to freeze meats, fruits, and vegetables but could not develop the market for them. It even lacked funding to develop new products. Bob was interested in selling fish sausage. He thought this was a good product because it could use fish species that were not highly valued, and the sausage could therefore be produced cheaply. But he suspended the project because he lacked the $3,000 necessary to develop sausage-making equipment. Bob had an idea for an improved freezer but did not have the money to develop it.

The company needed more capital to continue research and to develop new ideas. Otherwise it could never grow into a major industry. Banks would not finance a new frozen-food industry. They would finance the old bulk-freezing operations, but this new packaged frozen-food business seemed too risky and too improbable. Reminiscing in an industry magazine, *Refrigerating Engineering,* in 1953, Birdseye wrote, "A bank would no more loan on our inventory than on ice stacked in Death Valley!"

The best of the Clarence Birdseye stories is the one about the heiress and the goose. The heiress was Marjorie Merriweather Post, daughter of Charles William Post, who had made a fortune in Battle Creek, Michigan, building the Postum Cereal Company. When Post died in 1973, her estimated wealth was $200 million.

According to the story, one day in 1926 while vacationing on her yacht, she sailed into Gloucester Harbor. Her cook went into town for provisions and that night served her a goose. She was shocked to learn that this delicious goose, so tender, so, well, fresh tasting, had been frozen. Who froze this goose? How did they do it? She wanted to know. In the next scene in the story she is down at the Fort touring the General Seafoods plant. She then, according to the legend, spent the next three years trying to convince her husband—in some versions it is her father—to buy the Birdseye company, which he finally does for a record $22 million.

Americans love these rags-to-riches stories, like the girl being discovered at the lunch counter and becoming a movie star. Of course, in Birdseye's case, he didn't exactly come from rags. The story of the goose has often been told, in most accounts of Birdseye's life and in obituaries for Marjorie Post in the *Washington Post* and the *New York Times*, as well as in earlier books by this author. But upon close examination there is so little evidence and so many holes that it is questionable whether anything at all of the story is true.

We know that Marjorie Post did sail into Gloucester around 1926 because there is a photograph of her huge, four-masted bark in the harbor. The *Gloucester Daily Times*, in an August 15, 2005, article that demonstrated Gloucester's unique sense of which facts are important, completely accepted the story except for one detail: it was always said that she sailed in on her boat called the *Sea Cloud*. But the Gloucester newspaper pointed out that this vessel was not launched until 1931, so in all likelihood she actually sailed in on the *Hussar IV*, which was launched in 1923.

No one seems to question the goose, although in 1926 the only thing the appropriately named General Seafoods Corporation was selling was frozen haddock fillets. It was experimentally freezing many other foods, and that may have included a goose, though how the Post yacht would have gotten it is a mystery. Birdseye *might* have sent her the goose. At the time he was looking for investors, and he often sent prospects frozen food. One Wall Street investor was sent an entire multicourse dinner frozen. Most of the items were foods that most people had never heard of being frozen.

There is a faint whiff of sexism to the entire goose story. It ignores what an accomplished woman Post was. She definitely did not go to her father, because he unexpectedly committed suicide in 1914 and left her in charge of the company, which she ran for eight years. She was the principal shareholder. Her husband, Edward Francis Hutton, an aggressive financial tycoon who had founded the E. F. Hutton brokerage house before the marriage, took over the company in 1922. Still a young man, he was a legendary Wall Street operative who dazzled the financial world in 1906 by selling off his San Francisco investments by private telegraph before news of the great earthquake had reached the East. Marjorie Post, who had been well trained by her father, worked with her husband, her second of four, and together they acquired a dozen food companies. They rarely disagreed, at least publicly.

The problem with the goose story is that it portrays Marjorie Post as a self-indulgent heiress who on a whim, because she enjoyed her dinner, was prepared to spend millions—unless some man reined her in. But the truth was very different.

Marjorie Post's father, Charles Post, a man of Birdseye's father's generation always known as C.W., was born in Springfield, Illinois. He had a high-strung and restless nature and was given to depression, but he had a history of starting up new enterprises from ideas that excited him. Like Birdseye, once successful, he was eager to move on to the next idea. In 1890, he traveled with his wife and three-year-old daughter,

Marjorie, to Michigan, where he checked into a sanitarium under the direction of Dr. John Harvey Kellogg, famous for a treatment of "inspirational talks" and a grain-based vegetarian diet. Kellogg and his brother, Will K. Kellogg, concocted a line of health foods for their sanitarium, including "caramel coffee," a caffeine-free hot drink made from bread crusts, bran, and molasses.

C.W. emerged from the Kellogg diet weak and emaciated, and he finally restored his health with the Christian Scientists, who included a little meat in his diet. Still, Post was so taken with Dr. Kellogg's operation that in 1892 he opened his own sanitarium on a ten-acre farm outside Battle Creek and called it La Vita Inn. He was clearly copying the Kelloggs with a similar treatment and a special diet. He even had his own caramel coffee, also made with molasses and bran but with wheat berries. He published a book, *I Am Well!*, which promoted "mind-cure," a trendy idea popular with American businessmen that illness was not real, and the human mind could cure all. Dr. Kellogg too had been successful with several books on his health and diet ideas.

Neither Post's book nor his inn caught on. But he had an idea that he might be able to sell his coffee substitute. He started with mail order and in 1895 began selling it to grocers, branding the beverage Postum.

The first year Post lost $800 on Postum. That was when C.W. realized that his real talent was writing advertising copy. He launched a national campaign that offered to save people from the evils of coffee, which he claimed caused heart attacks, laziness, blindness, cowardice, and stupidity. Does coffee, he asked, "neutralize all of your efforts to gain money and fame"? A lot of people were not gaining the fame and money they hoped for, and quite a few of those drank coffee. Soon Post was becoming rich from Postum sales. In 1898, he came out with a cereal named Grape-Nuts. In the process of baking this concoction of whole wheat, yeast, and malted barley flour, starch turned into dextrose, the sugar found in grapes. And so they were "grape nuts."

Always an aggressive and inventive copywriter, Post claimed his products sent people "on the road to Wellville." In 1903 combined earnings from Postum and Grape-Nuts were more than $1 million and continued to grow; by the time Post died, he was one of the five largest advertisers in the country with an annual ad budget of $1 million. Post had eventually lost interest in most of the company's operations but continued to insist on writing all the advertising himself.

The Kelloggs followed Post's example. Will, the underpaid, mistreated kid brother with no medical degree, had the job of boiling wheat dough and pressing it into thin sheets in a roller. One morning in 1894 the dough was too dry and flaked off as it came out of the rollers. From that he developed a process called tempering, which produced flakes, and the concept of cereal flakes was born. He began experimenting with other grains. Corn was particularly popular. Will wanted to do what Post had done, but his brother was a man of science who felt that selling cereal would be beneath him. Finally, in 1906, Will broke away from his brother and formed his own cereal company, Kellogg's, which like Post's distinguished itself by its large and successful advertising campaigns.

Kellogg and Post were pioneers of modern marketing. But while Kellogg stuck close to its founding cereal products, Post was interested in most any new industrial food idea. To him making a new product take off was just a question of finding the right advertising campaign. Under his only daughter, Marjorie, and her husband, Edward Hutton, the company made regular acquisitions. In 1925 it bought Jell-O, an ambitious company that had expanded the market in flavored gelatin through marketing innovations such as distributing Jell-O recipe books. In 1926 it bought Baker's chocolate, one of America's oldest and biggest chocolate companies, and in 1928 it bought Maxwell House coffee.

By the time Marjorie Merriweather Post sailed into Gloucester Harbor, she was a savvy and accomplished businesswoman who had run a major corporation by herself and was in the process of buying out promising industrial food companies. She would have done more than just visit the plant; she would have found out everything she could about its business and its ideas and the value of the company and the ideas. Trained by C.W., she would have considered the potential for marketing and advertising. She would have talked it over with her husband, not because he was in charge, but because he was the one with the financial expertise. And then they would have gone to Wall Street to raise the money for the deal. That is exactly what happened. Whether a frozen goose was involved is not known. The origin of the goose story has been lost. Birdseye never told this story, which further suggests that it never happened. Instead, Bob always said that he and his partners had been trying to sell the company for some time, partly because they did not have the capital to develop their ideas and partly because the original plan had always been to make money by selling their ideas. Wetmore Hodges decided to go to people he knew at the Wall Street investment firm Goldman Sachs.

But Walter E. Sachs told the story differently. He later testified to a U.S. Senate committee that both Post and Hodges had approached him. His version did not include a goose. He testified that Post had wanted to buy the company and was looking for additional capital to be able to do this. In this version it was not doubts about frozen food from Hutton or the board of directors but a lack of sufficient funds to purchase the patents that was holding up the deal. Birdseye and his investors had always seen the patents as the real value, and Hutton seems to have agreed.

In 1929, Post did buy out Birdseye for a total of $23.5 million, according to the Senate investigation. Another well-worn legend is that Birdseye was trying to sell the company in 1926 for only $2 million, which was about the value of the company. So why did Post pay so much more? It paid over $20 million more because it believed in frozen food and wanted to own his patents. The often-mentioned $2 million price probably did not include the patents, only a small, not very successful seafood company.

The Birdseye company was of obvious appeal to the Post company. It had developed the ideas and technology for an entirely new food industry, but it lacked capital and had a huge image problem that would require skilled marketing and advertising. Each company had what the other lacked, and the two fit like sodium and chloride.

Birdseye always liked to boast that he received the most money ever paid for a patent—an impressive victory considering that most of his important patents had not yet been granted and wouldn't come through until 1930. Post bought all the patents for $20 million and then on top of that still bought the company for the remainder. The total $23.5 million deal was for the names, patents, patent applications, and all assets. Birdseye's personal share was about $1 million.

The acquisition was announced on May 7, 1929, with the Postum Company as the majority owner. On July 27 Postum stockholders approved the purchase. Postum reorganized its company, going on the New York Stock Exchange on July 25 for the first time under its new name, General Foods. It liked the Birdseye name and the Birdseye idea of becoming the leading food company.

Goldman Sachs purchased Postum stock to provide it with the capital for the acquisition. Postum took the $10.5 million it acquired from the stock sale and used it toward a controlling 51 percent of General Seafoods. Goldman Sachs spent $12.5 million on the remaining 49 percent.

Was too much paid? In the struggling 1930s the question was periodically asked. In 1939, *Fortune* magazine reported that in the first few years General Foods earned only two cents for every dollar it had spent acquiring the company. The outset of the Depression was not a good time to be building a new industry, but in April 1929 no one foresaw what was about to happen to the economy. Marjorie Post probably remembered that her father had lost money on his Postum beverage its first year too.

Goldman Sachs's eventual $12 million loss was the subject of a U.S. Senate Banking and Currency Committee investigation in 1932. Why, the committee members wanted to know, would anyone pay $23.5 million for a company valued at

$1.75 million. It was pointed out in the hearings that this represented a substantial loss for Goldman Sachs's stockholders. When a senator sneered that for its money all it got was a $1.75 million company, Sachs added, "And a process for frosted food that appeared to have great value." The ire of the senators and the stockholders arose from the fact that Goldman Sachs sold the Postum stock at a loss in 1931, selling its 49 percent back to the company for a fraction of its original value.

Goldman Sachs was the only loser. Post and Hutton got their company and with it fulfilled Birdseye's dream of a General Foods giant based on frozen food.

It was certainly a good deal for Bob Birdseye. The deal was signed in the spring of 1929, only months before the great market crash. A few months later General Seafoods probably would not have had a buyer or a future. While millions started the 1930s having lost everything, Birdseye went into the Depression with an extra million dollars. A million dollars in 1929 was worth $12 million in 2010 dollars. The top capital gains tax rate at the time was only 12.5 percent. In addition, he rode out the worst years of the Depression, continuing his frozen-food experiments with a comfortable salary as the director of research for General Foods' new Gloucester-based frozen-food division. The Birdseye mind was part of the deal. Unlike for most people, for Bob Birdseye life was looking good in the terrible fall of 1929.

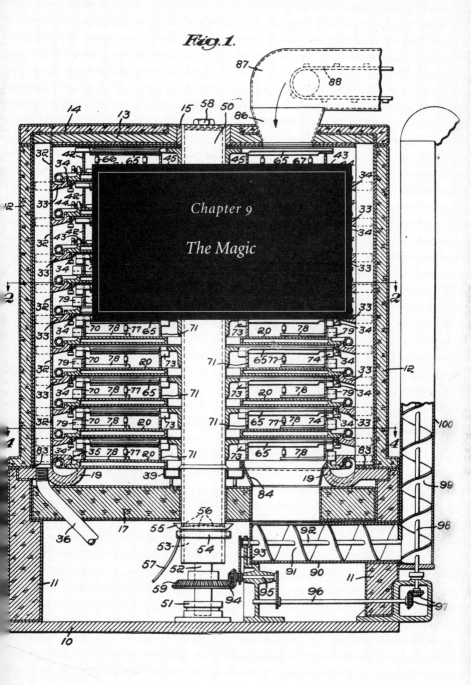

Fig.1.

Chapter 9

The Magic

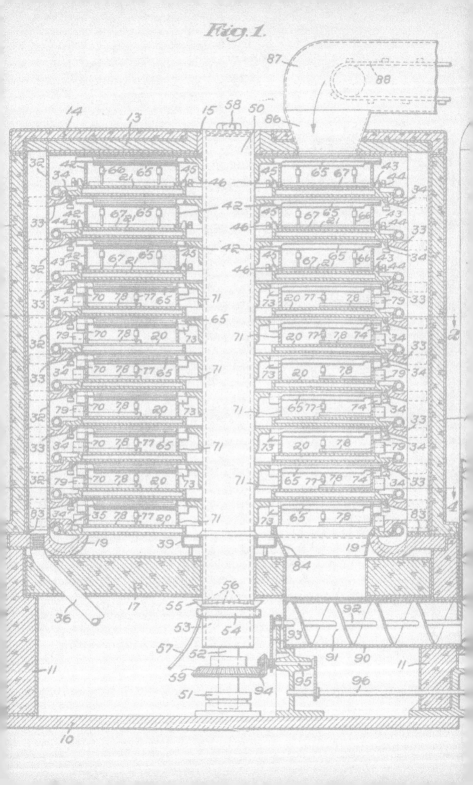

Fig.1.

Birdseye, the new millionaire, did have a few ideas about how to spend it. Paul G. Seyler, who worked for Thrift Stores in Springfield, Massachusetts, one of the first self-service chains, traveled to Gloucester to meet with Birdseye to discuss carrying his frozen food. Seyler noted that Birdseye drove up to the hotel for lunch in his Model A Ford but then phoned Eleanor to bring him the Packard 8. With its long, sleek body and high radiator grille crowned with a hood ornament like a trophy and the slogan "Ask the man who owns one," the Packard 8 was one of the premier luxury cars of the day. After lunch Birdseye took Seyler for a ride along the harbor to Eastern Point and showed him a long hill facing the opening of the harbor with a view out to sea. From his Packard, Birdseye took in the sparkling sea view and said to Seyler, "All my life I have desired a Packard 8 car and that property. Now I have them." Seyler later said, "Despite that evidence of materialism I was convinced from my conversations with Mr. Birdseye that he was an inventive genius more interested in contriving something beautiful and useful for humanity than in monetary gain."

Seyler may have been right, but still Birdseye's comment was telling. Birdseye was in his mid-forties, and all his life he and his family had teeter-tottered between privilege and ruin. Now he was rich.

He did not have the eccentricity of a Hammond to build himself a medieval castle. To him a baronial mansion seemed more appropriate. He had his seventeen-room mansion built in an astounding seven months between December 1930 and the end of June 1931. When completed, it was valued at $200,000, which would be worth $2.7 million today, an impressive expenditure for the Depression.

He tirelessly shot 16-millimeter black-and-white film and documented every step in the construction of his house from excavating the foundation with shovels and pickaxes to planting shrubbery and furniture being carried in on moving day. He also shot film of his children, his dogs, snow blizzards, a whale spouting, the 1931 solar eclipse, and some of the Gloucester sights that drew so many artists: fiery sunsets, storm surges crashing waves on the rough granite coast, the tall-masted fishing fleet in Gloucester Harbor.

The furnishings betrayed no interest in the design fashions of the 1920s and 1930s, with a lot of wooden paneling and trim, mahogany and marble mantels, dark wood furniture with floral upholstery, and Oriental rugs. Birdseye imported delicate floral wallpaper from France for the dining room. There was a screened porch off the dining room, a pine-paneled study, a sunporch with a wall fountain. The five spacious bedrooms upstairs afforded stunning views of Gloucester Harbor and out to the open bay and the Atlantic. The artwork was of no real interest, which is striking because in Gloucester, even today, but especially in that period, many people collected the excellent locally produced art.

There was a shooting range in the basement, though the basement ended up being used more as a laboratory for Birdseye's experiments. The closest the house came to state of the art was, not surprisingly, the kitchen. Don Wonson, from an old Gloucester family who grew up next door to the Birdseyes, recalled the unusual kitchen:

> The kitchen was all stainless steel. He was very meticulous about the kitchen, wiping it down with towels—stainless steel sinks, big refrigerators, big stainless steel with double doors. A big freezer in the kitchen. More refrigerators downstairs. Stainless steel counters, even the drain boards were stainless steel.

Birdseye handpicked every detail. In an age when there were no blacktop driveways, he put in an innovative tar-and-crushed-granite pavement. The garage, which housed their three cars, was so large that the top floor was a spacious apartment. The huge wooden doors to the garage resembled gates to a castle, with massive iron strap hinges.

For people known for their simplicity and straightforwardness, it was an imposing mansion on a hill. Four tall white pillars announced the front entrance. And of course he named the house Wyndiecote and finally had a home worthy of the name. Perhaps nothing better illustrated who Bob Birdseye was than the fact that the Packard he took so much pride in owning usually stayed in the garage. Bob was almost always seen in his little Ford.

One old friend from the neighborhood, Lila Monell's sister Josephine, said, "They worked and lived under miserable conditions. All of a sudden, money at last. So let's buy a house. They didn't put on airs—except the house. But what do you do if you suddenly have a million dollars? Let's build a house!"

In Eastern Point they had chosen an unusual setting. The

narrow peninsula at the opening of Gloucester Harbor, with the harbor on one side and the rugged granite Atlantic coast on the other, offered stunning views. But it had never been fully developed. In Gloucester people with money had always preferred to live back from the fishy waterfront. Until 1889, when Gloucester was already almost three centuries old, Eastern Point was simply farmland. But that year a developer planned to cover it with houses, large, stately houses with spectacular ocean views. He built the first eleven houses but then went bankrupt trying to build Eastern Point's first road. When the Birdseyes built their house, it was one of the first on Eastern Point to be fully weatherized for year-round occupancy.

Rather than an arrogant expression of his new position, the Birdseye house reveals a certain system of beliefs with which Bob had been raised. This brick house with white pillars perched on a hill did not look like other Eastern Point homes, or really other Gloucester mansions. In spirit it resembled the large, solid, wooden eighteenth-century houses built by shipowners and cod merchants, placed on the high ground over downtown Gloucester to look down at the harbor from where their wealth came. Like the old shipowners, Birdseye had built a house that stood as a symbol of economic success, of the solidity of American capitalism, of the reassuring power and stability of money. Birdseye thought like this, and at the outset of the Depression the many people who were starting to lose their faith in this outlook could look up at the Birdseye temple and see that at least capitalism had worked for someone. Then too it had a certain old-fashioned grand bourgeois feel like the brownstone where he was raised in Cobble Hill, Brooklyn.

. . .

The original plant manager, Joe Guinane, said of Birdseye, "He was an impatient man and had little interest in the routine affairs involved in running the business. Once the problem was solved, he was eager to turn to something new." Now that he had sold his frozen-food idea, he might have been eager to move on. Some even suggested that he might retire. After all, he was in his mid-forties and had made his fortune. But Birdseye was a restless man who probably never would have retired no matter what age he reached. He said, "Following one's curiosity is much more fun than taking things easy."

He was not yet through with frozen food. The problem had not yet been solved. It was still a long way from a real industry and an accepted part of American shopping and eating. All he had accomplished was getting the right people and money behind the project.

Birdseye essentially continued his work in Gloucester, only now as a salaried executive for the Gloucester-based Birds Eye Frosted Foods division of General Foods. Breaking up the name "Birdseye" is often said to have been based on giving the product greater market appeal, but there is also the legal issue that proper names are not supposed to be used as trademarks. One of the early marketing decisions, along with breaking his name into two words, was calling the product "Frosted Foods" to emphasize that it was something completely different from the frozen food that people knew.

Birdseye and General Foods were venturing into new territory. Birdseye later wrote, "Quick freezing was conceived, born, and nourished on a strange combination of ingenuity, stick-to-itiveness, sweat, and good luck."

By the fall of 1929 the new Birds Eye company was operating at capacity, stockpiling frozen food for its launch. On

March 6, 1930, the local newspaper in Springfield, Massachusetts, ran an advertisement with the headline "The Most Revolutionary Idea in the History of Food Will Be Revealed in Springfield Today."

The vaunted marketing and advertising experts of the former Post company had chosen Springfield as their site for selling twenty-seven different frozen items, including porterhouse steak, spring lamb chops, sliced ham, pork sausage, June peas, spinach, Oregon cherries, loganberries, red raspberries, fillets of haddock or sole, and bluepoint oysters. Twelve "demonstrators" were sent to ten participating stores. The demonstrators, all of whom were women, as were most of the shoppers, and the store staffs had all been put through a three-day training program largely focused on how to explain the difference between slow and fast freezing.

The stores were given display freezers worth $1,500, which was far more expensive than most of these small family stores could have been able to buy. The contents, the various frozen foods, were sold on consignment. The freezers were far from flawless, and a team of mechanics, assigned to the ten stores, was constantly checking temperatures and making repairs and adjustments.

The ad, which featured the new label, an upside-down bird and the name Birds Eye Frosted Foods, claimed it was "little short of magic." The haddock, it boasted, was "as fresh flavored as the day the fish was drawn from the cold blue waters of the North Atlantic." And "Here is the most wondrous magic of all! June peas, as gloriously green as any you will see next summer. Red raspberries, plump and tender and deliciously flavored. Big, smiling pie cherries—and loganberries. Imagine having them all summer-fresh in March!"

This kind of hyperbole continued for many years, not only from the Birds Eye company, but also in major newspapers and magazines. In February 1932 the *New York Times* called frozen food a "scientific miracle in home management." It reported on such items as peaches, strawberries, and oysters becoming available out of season, as though a frozen berry and a fresh one were indistinguishable. *Popular Science Monthly* claimed that frozen food and fresh food were "exactly the same." Today's discerning palate knows that frozen and fresh are considerably different, but for people who were accustomed to canned food or slow-frozen food, the new "frosted food" did seem miraculous. Consumers who had the courage to try the new frosted food were pleased with how much better it was than they had expected. When frozen fish was test-marketed in New England, three out of four purchasers of frozen fillets came back to buy more.

While there was a positive initial response, frosted food was not an instant success. One of its drawbacks was that a major advantage of frozen food, that you could buy it now and use it whenever you wanted, was not yet a reality since few people had adequate home freezers.

The owners of the first ten stores were people who believed frozen food was the future and were willing to proselytize to a wary public. Seyler said, "I clearly recall that it took five minutes of fast talking to sell a reluctant housewife one thirty-five cent package of Birds Eye frozen peas." According to Seyler, "The consumer just couldn't understand how anything frozen could possibly be safe, let alone good to eat."

One of the store owners, Joshua Davidson, who had taken over his father's business, in an interview in *Quick Frozen Foods* magazine thirty years later, recalled, "I could see this as a sound venture, a progressive step in the food industry and I wanted to be in step. That's why I agreed to be one of the first ten dealers." In truth General Foods, in supplying the freezer and letting the store have the food on consignment, had made it risk free for the store owner. "Frankly," said Davidson, "I couldn't afford not to be the way Birds Eye offered to start me."

But to make it work, Birds Eye had to find stores that were deeply committed. Davidson said, "The food customer of 1930 was prejudiced against frozen foods. She felt foods were frozen because they were low-grade or spoiled and had to be frozen to be salvaged. So we had to convince the people frozen foods were okay. It was a mission and we had to do a lot of talking."

Davidson said that he was constantly answering questions about how the food was made, how to cook it. One woman asked if the company would start packaging meat and vegetables together. Davidson dismissed that idea, pointing out that people would prefer to choose their menus themselves. It took twenty years, but frozen TV dinners became a popular item of the 1950s, almost a symbol of a peculiar suburban affluence of the era.

Mostly, people wanted to know if the frozen food had any taste. Many people asked questions but would not buy. But after a month of answering questions, Davidson saw the amount of frozen food sold in his store double.

But frozen food was slow to catch on. One of the problems was that fresh food was relatively inexpensive. A top-quality steak often sold for less than thirty cents. It was inevitable that frozen food had to cost a few cents more than the fresh food it was made from. In the first decade of the Depression, canned food was for the poor, fresh for the middle class and affluent, and out-of-season frozen food was a high-quality luxury for the wealthy. Only in the 1940s, when Birds Eye started getting competitors, did prices come down and frozen food become popular with middle- and lower-income people.

In 1930, Birdseye invented a portable multiplate freezer. A typical Birdseye invention, it was built from scraps of corrugated metal with steel plates and coils carrying refrigerant. It could be brought out to fields to freeze produce as it was harvested.

But though such devices increased production capacity, it was for a future that had not arrived. Retail interest in frosted food was not expanding along with capacity. By 1933 there were still only 516 retail stores in America carrying frosted food. When the company questioned retailers, the chief complaint was the store freezer units. Though they were given to the stores, they used a great deal of electricity and were expensive to operate. Birds Eye was not happy about the freezer either, because giving away $1,500 machines was a losing proposition.

The Frozen Foods division was losing money, and General Foods put Edwin Gibson, a former mortgage banker, in charge of making the division profitable. Gibson, operating out of New York, saw the freezer units as the first problem. Birdseye, who had been in discussions with the American Radiator Company through Wetmore Hodges since before the Post buyout, agreed. Gibson gave a priority to the freezer American Radiator was developing. The Amrad cabinet turned out to be a huge step forward for frozen food. It put the frozen food on display with a slanted window in the front, but, more important, a freezer cost only $300, which meant the food could be sold at an affordable price without General

Foods losing money on the cost of the freezers. And stores could afford to use them because the running cost for the new freezer was about $3.50 a month instead of the $16.00 and $20.00 per month of the old freezers. General Foods rented the Amrad units to stores for $10.00 to $12.50 per month. In 1934, Gibson ordered a new Springfield-style market test with the new freezers in Syracuse, New York. The freezer did so well there that another test was launched in Rochester. From there they spread throughout the Northeast into the Midwest and were well on their way to national distribution.

Birdseye himself used his talents for persuasion, traveling to stores to hand sell his product. "He was one of the most articulate and persuasive men I have ever known," Joe Guinane said. He had also become a minor celebrity. People in the Depression were hungry for success stories, and they liked the tale of the curious little genius—he looked the role—who through his remarkable resourcefulness had made a fortune on frozen food. He was booked for events that advertised, "See a demonstration of quick frosting by Mr. Clarence Birdseye, the inventor of the famous quick freezing process." Sometimes the public was also invited to "meet Mrs. Clarence Birdseye." Mostly women attended these demonstrations, but at one event in a Boston hotel, Bob's old friend from Labrador Sir Wilfred Grenfell came to see the Birdseye demonstration.

Even if the Frosted Foods division was struggling to make a profit, there seemed to have been no limit to the hyperbole directed toward frozen food and even Birdseye by journalists, politicians, and businessmen. In 1931 the mayor of Boston, James Michael Curley, hosted a dinner to honor Birdseye, according to press reports calling him a genius "whose contribution to the welfare of mankind gives promise of being the greatest in volume and value in a half century of American history." After Birdseye gave a demonstration, freezing a steak with dry ice, the mayor stated that Birdseye's invention "will probably be a greater contributing factor in preventing wars in the future than battleships or any other agency, because, after all, wars are caused by starvation."

On his $50,000 salary, Birdseye was still pondering the mysteries of freezing. He wanted to try freezing everything, including prepared foods, and he never tired of experimenting with new frozen species such as porpoise and whale. As though he were still living in Labrador struggling to procure fresh food, he ate whatever wildlife he found in Gloucester, including birds he trapped. He would freeze these catches to see if they froze well. He was particularly fond of coot, a variety of waterfowl.

He also liked to investigate more abstract concepts of freezing. How long would frozen meat keep? Forever, said scientists, if it was not exposed to air, though some speculated that frozen animal fat would deteriorate after a few centuries. Birdseye reflected on Eleanor's moose. It had become legendary in the Birdseye family that in 1929 Eleanor had shot a moose, though where she had hunted it has been lost in time. They butchered it and froze it, thawing out a part from time to time for dinner. By 1933 nothing was left but the neck. Even the omnivorous Birdseye did not like moose neck because it was tough and gamy. Would four years of freezing change it? He had to find out, so he thawed it and cooked it, and to his surprise found it to be tender.

To further his research, Birdseye, who always sought to surround himself with highly qualified people, hired a chemist experienced in food issues, Donald Tressler, to head the research team. Birdseye arranged for Tressler and his family to move into his old home with the long shady porch on Beach Road. Birdseye and Tressler conducted experiments measuring meat tenderness with a tire pressure gauge and a laboratory instrument called a penetrometer. They demonstrated that meat became tenderer after a week of quick-freezing than before it was frozen.

Birdseye still wanted to find out how the Inuit fish had survived freezing. They had frozen fish in the air, and the fish had still been alive months later when thawed. How can a living organism survive freezing? The question was more than a century old. In the late eighteenth and early nineteenth centuries, when scientists started experimenting with faster ways of freezing and arriving at ever-colder temperatures, the question periodically arose whether an organism, if frozen fast enough, could have its functions slowed but not stopped so that it could be restored on thawing. Bob, who swore he had seen this accomplished by the Inuit in Labrador, would catch pickerel in the nearby Niles Pond, and try to freeze them and keep them alive. Birdseye's eldest son, Kellogg, always remembered his mother's irritation at continually finding fish flipping around in the bathtub. Bob would explain to children in the neighborhood what he was trying to do, and it was generally assumed that it could not be done, that Mr. Birdseye was brilliant but a bit eccentric, though some thought that if Birdseye believed it was possible, it was.

At the exact same time, 1934–36, scientists in the Soviet Union were wondering about the same thing. They concluded that freezing killed animals. However, they noted that an exception was fish. They demonstrated that when ice formed around the skin and the subcutaneous tissue directly under it, the fish would become hard and appear to be frozen, but the living organs inside would not freeze and could continue to function. In 1934, N. A. Borodin in his study "The Anabiosis or Phenomenon of Resuscitation of Fishes After Being Frozen" concluded that it could be done and that arctic species were particularly suited to survive freezing. The fish Birdseye had observed in Labrador were pulled into −40-degree air, and the water on them instantly froze, possibly leaving the interior of the fish unfrozen and unharmed.

At the exact same time, 1935, a man whose name and profession have been lost walked into the office of E. W. Williams, who had recently become the publisher of a magazine on the meat trade, *Butchers' Advocate*. The visitor offered to demonstrate something he called suspended animation. He showed Williams a goldfish in a bowl. He took the fish out and plunged it into dry ice, freezing it instantly. The fish was rock hard and lifeless when he dropped it back in the bowl. It sank to the bottom. After a minute the tail started to wriggle. Soon it was swimming circles around the bowl. Shortly after that Williams saw a demonstration of quick freezing by the Birds Eye company at a food fair. Years later he would cite the combination of Birds Eye's out-of-season fruit and the revived goldfish with the intriguing label "suspended animation"—and as the two experiences that made him think frozen food was "an industry with a future." In 1938 he started an industry magazine called *Quick Frozen Foods*, on which Birdseye served as an adviser for the rest of his life. The magazine now has a Web site and continues to report on the frozen-food industry. As for suspended animation, it has led to the science of cryonics, a controversial field in which humans are frozen shortly after being pronounced legally dead in the hopes that a still nonexistent procedure will restore them to life at some time in the future.

When Birdseye had an idea, no matter how nutty it seemed, it was always worth considering. Some of his ideas seemed to be great successes, even though they were never used, such as a machine he built to freeze vegetables individually, rather than in a block. A pea would come down a chute and land on one of a series of revolving freezing plates.

Though his laboratory was small and only employed twenty-two chemists and assistants, Birdseye was fondly remembered in Gloucester as someone who offered work through the Depression. In a 1980 article in the *Gloucester Daily Times*, Bill Nickerson, a former Birdseye worker, said, "There was nothing as far as jobs go then. Mr. Birdseye gave me a job right out of high school, when I couldn't find another one." He worked as a thirty-cent-an-hour laboratory assistant.

In this laboratory the team worked out the problems of freezing vegetables, fruits, and prepared foods. They learned that vegetables degenerated due to an enzyme that could be removed by blanching—a quick plunge in boiling water. Then the vegetables were rapidly cooled, quick-frozen, and packed in watertight cellophane. This left them with a bright color. Frozen peas became one of the most successful Birds Eye frosted foods because they were such a brilliant green.

The laboratory team also solved the problem of frozen sliced onions, which always turned black with freezing. They found that if blanched for just the right length of time, the onions would remain white. Fruits were difficult because they softened and oxidized brown with freezing. The oxidation was solved with a small amount of sulfur dioxide or ascorbic acid, which is vitamin C. But they were not able to completely solve the problem of texture. An expert from Cornell, Lucy Kimball, was brought in to develop techniques for cooking prepared foods so that after being thawed and reheated they maintained their quality.

As the market for frozen food grew, variety became the new problem. Retailers were complaining that there was not a wide enough selection of products. Birdseye established a partnership with a large farm in southern New Jersey, Seabrook Farms, to supply a greater variety of vegetables. Birdseye had been encouraging farms to become industrial, but Seabrook was already an agro-industrial operation, supplying produce for canneries. In 1911, Charles Seabrook had started Seabrook Farms as a model of modern agribusiness. An engineer, he built power plants and irrigation systems, so Birds Eye was able to have them wash, blanch, and cool produce for freezing in the new portable freezers. Lima beans, which had been a mainstay of the farm's cannery business, were one of the early frozen-vegetable successes from Seabrook.

Certain problem products were never conquered. Birdseye could not freeze lettuce without wilting it or tomatoes without ruining their texture, and frozen bananas were a disaster. But between 1932 and 1934 more than one hundred types of frozen foods, many of them processed foods, were developed. Most of them were not put on the market until the 1940s because of the lack of infrastructure and consumer interest.

Tressler, the head of research, wrote about 1929, when the laboratory began its work, "This was the beginning of the depression but we did not realize it." By 1933 General Foods realized it. It still believed in the future of frozen food and continued to aggressively market it, but the company had to cut its losses. It closed down the Gloucester operation, worked out of Boston, and did not return to Gloucester until the 1940s.

Bob Birdseye went from president of the division to consultant. He continued to take an interest in the progress of frozen food and to promote it whenever he got a chance, whenever he was asked, sometimes without being asked. But his restless mind moved on to other ideas.

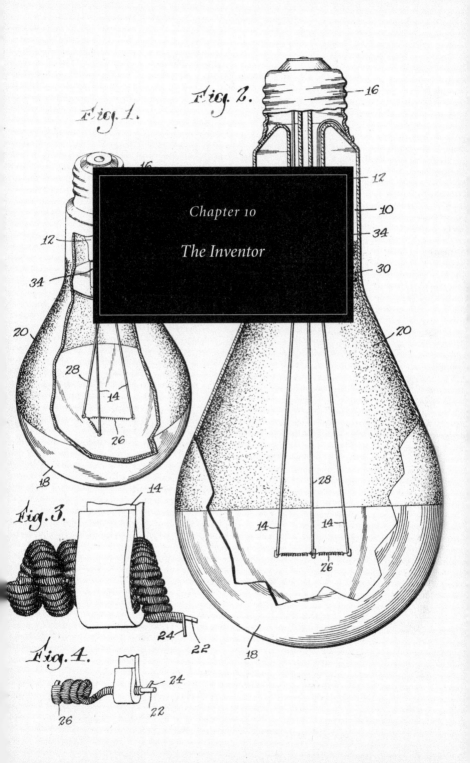

Fig. 1.

Fig. 2.

Fig. 3.

Fig. 4.

Chapter 10

The Inventor

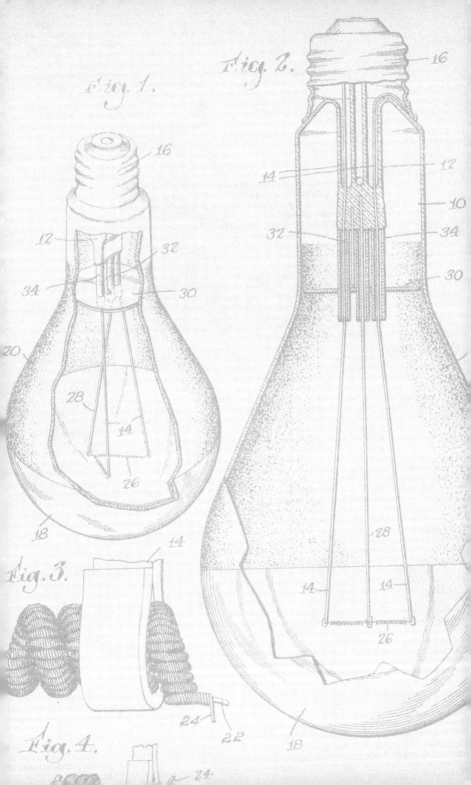

Fig. 1.

Fig. 2.

Fig. 3.

Fig. 4.

Safe from the Depression, the Birdseyes had found the good life in their Eastern Point Wyndiecote. Don Wonson remembered the neighbors of his boyhood: "The house was like a showcase. Andy the gardener was always calling people over to see something blooming in the garden, fish pond with pussy willows, pond stocked with carp, goldfish, and some unusual fish and aquatic plants."

As soon as the Eastern Point house was completed, the Birdseyes built horse stables in West Gloucester. Gloucester is in effect an island cut off from Massachusetts because the strip of land connecting it, only a few yards wide, was dug through in the seventeenth century, enabling ships to sail from the north coast of Cape Ann to Boston without rounding the treacherous rocky tip off the cape. The town grew upon this island, centered on the perfect harbor. West Gloucester, the side on the mainland, was a more rural area where the wealthy retreated from the redolent drying fish in the hot summer months. There the Birdseyes owned five hundred acres of land and built not a little barn for horses but an elegant two-story riding stable with a long arcade. They kept fine Thoroughbred horses and rode with slim English saddles. Birdseye, who had lived on horseback in the Southwest and in the Bitterroot, was an expert horseman, and he taught his children. They galloped through the woods, went jumping, and even did some stunt riding. Bob liked to shoot handguns while mounted, which took some skill since the horse would lurch, alarmed by the pop of the pistol. They would often ride to Wingaersheek Beach, where Bob liked to ride up the sand dunes and get his horse to slide down on its backside.

Between the Eastern Point mansion and the West Glouces-ter stables, Birdseye had reproduced his childhood with the elegant Cobble Hill brownstone and the farm on the North Fork of Long Island.

The Birdseyes had bird dogs, Mickey the Irish setter, who greeted everyone but had to be muzzled because of his bad habit of nipping, and Jeff the sweet-natured Brittany. Bob took his dogs duck hunting in South Dakota every year. He loved hunting and taught the whole family to shoot, sometimes tak-ing them with him on hunting trips. He was an outdoorsman. At first he looked too small and mild for the part, but then there were those large hands and the weather-beaten face.

The Birdseyes joined the snobbish Eastern Point Yacht Club, and the children learned sailing. They frequently enter-tained. Bob was remembered as extremely open and casual, Eleanor as a little more formal and reserved. Bob loved to cook, and they had dinner parties in their large formal dining room with thrilling sea views. He specialized in what he liked to call "lobster feeds," during which he would triumphantly emerge from the kitchen sporting an apron and chef's toque and carrying an enormous platter of steamed lobsters.

Judging from the amount of film he shot of sunsets over the harbor and whitecaps spraying over the dark crests of the granite boulders, Bob must have loved these views. He seemed to love Gloucester. He cooked seawater down to salt and carried that salt in a vial wherever he traveled. Despite the stiffness of his writing, Bob was a sentimental man.

He tried to take care of Eastern Point, preserving its bird habitat. He persuaded the government to build a seawall in a nearby cove to protect Niles Pond from storms. He raised animals behind his house. He had chicken coops with long runs for Rhode Island Reds, the same species he and Eleanor had raised in Labrador. They also raised big fluffy chinchillas. Bob would sell the pelts and freeze the meat.

Buffalo Bill's Wild West was a never-forgotten romance. Bob was still an avid reader of Henty adventure novels and kept rereading his favorite, *Redskin and Cowboy*, as well as the western novels of Owen Wister, famous for his 1902 *The Virginian*, and the western stories of Bret Harte. He also enjoyed going to movies to see westerns or staying home to play Chinese checkers, a game for which he had a near-fanatical passion.

Bob was a curious figure in Gloucester, famously an inventive genius who did odd things that regular people would not understand. When rats invaded his melon patch, he stood on the terrace and shot them with a handgun. He was always trapping or shooting or freezing something as though he were still surviving in a frontier. Children would ask him what he was doing, and he would always eagerly explain. He would even take them to his kitchen or his basement to show them an experiment he was working on. "If you were curious, he would explain how and why and when. He took the time," said Don Wonson. This was a lot more exciting for local kids than a man giving out candy.

His daughter Eleanor, when interviewed at age eighty-two

in 2005, recalled early childhood with Dad, learning how to dissect animals in the Persian-carpeted living room. She also learned how to skin mice and how to cure skins. He would often involve his own children in his experiments. Sometimes the experiments were for no particular purpose other than to satisfy his own curiosity, such as when he got an idea for a new way to make potato chips and enlisted the family in potato-chip making. His wife, Eleanor, too, when she did not run out of patience, was enlisted in kitchen experiments.

While his kids were learning to be yachtsmen at the Eastern Point club, he preferred to go fishing on his rugged forty-foot, wooden-hulled, open-deck power launch, the *Sealoafer*. If he caught something interesting, he would try freezing it. He saw whales and dolphins, and after trying to freeze them too, he started to become interested in them. As was the family tradition, he had someone film him at sea. There was the small bald man with glasses, dressed in a sweater and dress shirt and tie, walking out on a narrow plank over the bow of his boat with an enormous harpoon, which, like Queequeg in *Moby-Dick*, he would raise up and hurl into the side of a six-foot shark or a porpoise or a whale.

Was he wearing the tie for the camera or the fish? Birdseye never lost the family habit of dressing for a law office. What he would do with these animals is not certain, but as his neighbor Lila Monell pointed out with the starlings, when Bob got an animal, it usually ended up frozen or eaten, or both. Caught in 16-millimeter black and white is the same Bob Birdseye of the Bitterroot, a relentless predator. In one sequence he is harpooning porpoises, and then there is a scene with a dozen or more lying on the beach, some wriggling, some raising their heads and opening their mouths as though to cry out. Hunting marine mammals was legal in the 1930s.

According to the Birds Eye company official history, he started working with the International Whaling Commission, a group that tried to monitor whale migration to promote more rational hunting of them. But that organization did not exist until after World War II, and he was whaling from 1934 to 1938. Supposedly he was tagging finbacks, after the blue whale the second-largest animal that has ever lived on earth, with a heavy four-inch aluminum harpoon of his own invention. Aside from this machine, which still exists in the collection of the Peabody Essex Museum in Salem, there is disturbingly little evidence that he was tagging and not just killing. To insert the harpoon, he invented and patented the kickless, handheld whale harpoon. Typical of Birdseye, it looks as if it was handmade in a basement, which it probably was, and combines medieval technology with pragmatic inventiveness to solve his problem, an effective way to single-handedly plant the harpoon. The invention owes a lot to the crossbow. It has an aluminum rifle stock and a thick tube of black rubber to propel the harpoons. The rubber is stretched back by the turning of a large steel cogwheel with a hand crank, and a trigger releases it to fire the harpoon. It was much more efficient than throwing the harpoon spear-like, although Birdseye seemed effective enough with the old technique. Limited in its applications, the tool never caught on, and there is a record of only one ever being built. But it served him well at sea for four years. According to Bob, he harpooned fifty-two finbacks. There is also film footage of him using the harpoon gun to hunt sharks, and he may have found other uses. In the film footage he appears to be killing and not tagging.

He had a number of ideas that were not developed. One winter he was in Galveston, Texas, and he watched the commercial fishermen landing red snapper. They were hand lining, dropping a baited and weighted line many fathoms to the coral reefs on the ocean bottom. Then, when they got a bite, they quickly reeled the line up on their circling two thumbs the way hand liners have been fishing for centuries. Birdseye thought that there must be an easier way to do this. But when he talked to fishermen, they told him, the way fishermen often do, that they had always fished this way and it worked, so they were not interested in changing.

This was absolutely contrary to the Birdseye creed. Birdseye always said, "Just because something has always been done in a certain way is never a sufficient reason for continuing to do it in that way." Birdseye believed in change, believed in the constant updating and improvement of ideas. "Change," he asserted in his 1951 *American Magazine* article, "is the very essence of American life." And he frequently stated, "There is always a better way of doing almost everything. Today anything which is twenty years old is, or should be, apt to be obsolete."

So despite the protests of fishermen, Birdseye worked on the automatic reel: a device installed on the side of a boat that lowered a baited steel cable to depths greater than one hundred fathoms. When a fish bit, it set the hook and hauled it up to the surface, and even landed the fish on the deck.

Birdseye boasted, "The gadget does everything, in fact, except mix a mint julep for the fisherman." Birdseye thought he was onto the next big idea. He thought it would revolutionize commercial fishing because in one day the device could catch as many fish as three fishermen. He predicted that it would "make more sea food available, and increase fishermen's earnings." Like most people in the 1930s, Birdseye had no concept of overfishing. He never developed the device, and no one ever used it. Had it become a tool to revolutionize commercial fishing, its destructive power would be unacceptable in today's overfished ocean, and doubtless some would be calling for it to be banned.

Of greater success and far more impact was his work with lightbulbs. Birdseye noticed that reflectors were placed behind lightbulbs to illuminate displays in shopwindows. "I didn't know the difference between an ohm and a kilowatt," he said. "But it seemed to me there was no reason why the bulb and the reflector should be separate units. Wouldn't it be simpler and cheaper to build a lamp which would contain its own reflector?"

And so he formed the Birdseye Electric Company, a lightbulb company providing more jobs in Gloucester. His 1935 patent 2,219,510 for a reflecting electric lamp was one of his most important. Designed for working areas and window displays, his lightbulb with the built-in reflector has become a fixed idea in lighting. He also designed more intensely glowing filaments for more efficient lighting, and heat lamps for keeping food warm, another idea still in use. He also manufactured neon bulbs in the shape of the old radio tubes, with a decorative figure inside that would light up—flowers, animals, religious symbols, and occasional advertising such as his bulb in which the RCA logo lights up—all rare and very collectible today.

Birdseye developed ideas and used them to create a company, and then he would sell the company. That was what he did with frozen food, and in 1939 he sold the Birdseye Electric Company to the Wabash Appliance Corporation of Brooklyn on the condition that Birdseye Electric would keep its own name and personnel. Birdseye and Wabash continued until 1945, when Sylvania bought them out and continued producing Birdseye bulbs.

After several years of flooding the U.S. patent office with lightbulb inventions, Birdseye returned to freezing. In 1939 he came out with an important new invention, the gravity froster. It was to be his last important contribution to frozen-food technology. The gravity froster, in one sense, was a complete departure for him. While his idea had always been to freeze food in convenient retail packages, this freezer was for individual pieces of food that the producer could then package. This was an industrial freezer that did not take up a large amount of floor space in the plant, was portable, and could freeze loose products without drying them out. Loose vegetables were fed into the top of the froster and sent around originally ten extremely thin stainless steel plates—by 1941 twenty plates—with a liquid ammonia refrigerant. Scrapers removed frost from the plates and deposited the moisture on the vegetables until the pieces came out frozen at the bottom. The machine had enormous capacity, and one worker could operate four machines at once, because the produce moved through the machine automatically. A twenty-plate freezer could freeze eighteen hundred pounds of fresh peas in an hour. It could be used anywhere because of its excellent insulation. Previous freezers needed to operate in a cold room.

The machine was tested in the summer of 1939 in Gloucester, freezing peas, lima beans, oysters, and sliced strawberries. Then a Boston-based consortium of New England investors, Gravity Froster Corporation, was formed and leased the machine to an ever-expanding list of frozen-food companies.

In 1930, eighty thousand pounds of food were sold frozen. By the mid-1940s, ten times that much was being sold every year. Frozen food grew, as Birdseye always said it would, from a curiosity to a major industry.

There was still the problem of infrastructure. The industry was struggling to get enough well-insulated refrigerator trucks for shipping to its growing market. And there was still the problem of image. Birdseye and other advocates of frozen food were a major force in pushing Congress into its 1938 revision of the 1906 Pure Food and Drug Act. Birdseye felt that the standards of the 1906 law were not nearly stringent enough for the frozen-food industry. From the outset frozen food had suffered from a reputation for low quality. Birds Eye had introduced high standards, and it was thought that the only way the industry could grow, as new frozen-food companies were being created, was to make sure that all frozen food had these high standards so that the public would start associating frozen with quality. Birds Eye now had competing frozen-food companies, and many of them were willing to offer inferior products to keep the price down. This caused Birds Eye to lower its prices and mass market frozen food.

But Bob always understood the risk of lowering quality to lower prices. The 1938 law increased penalties, had a longer list of harmful commodities, had tougher rules about mislabeling and adulterating, and required a list of ingredients on every product's label. It has remained the underpinning of product safety in the United States.

There were still resisters, those who insisted frozen food was inferior. And some old arguments persisted.

Birdseye realized that after the war the opportunities for frozen food would be tremendous. World War II ended the resistance. A shortage of metals led to a decline in canning, which caused many people to try frozen food for the first time.

The war changed the way Americans ate, and the biggest difference in American eating was that during the war women left the kitchen for wartime jobs to do their part. Many were not going to go back or at least not full-time. And so a trend began in the 1940s that has continued ever since of looking for easier and quicker ways to prepare meals. In 1951 a survey in *Science Digest* found that 41 percent of American housewives preferred frozen food to either fresh or canned because of its convenience. Another huge change was the growth of supermarkets, which gave a great deal of their ample space to frozen food. By 1950, according to the American Frozen Food Institute, 64 percent of American retail food stores carried some frozen food.

But for Birdseye it was not enough to be the guru of the last big food idea, he wanted to be master of the next. Birdseye was convinced that the next important food idea was dehydration. He wondered if dehydrating—extracting water—more rapidly would have a similar result to freezing rapidly.

He began his work on dehydration in his kitchen with a coffee hotplate hung upside down from the ceiling with a plate of bread cubes placed a few inches below. Then he brought in a favorite piece of Birdseye equipment, an electric fan, to blow on the bread. With both the heat and the air blowing, he stirred the bread with a spoon. The combination of heat, stirring, and air, he concluded, dehydrated faster than the standard technique of blowing hot air. There were other techniques. Pouring milk on a heated stainless steel drum dried it. Fish meal was produced, and dried beef for World War II soldiers was made by placing the food inside a heated drum.

But all of these techniques resulted in losing flavor and nutrients. There were three known ways of transferring heat—convection, conduction, and radiation. Traditional dehydration had used only one, convection, the transfer of heat by moving warmed matter, in this case air. But there were also conduction—the movement of heat from atom to atom, thereby moving heat through the matter that is being warmed; and radiation—heating through heat waves such as the sun gives off, or in Birdseye's case from an infrared bulb. Birdseye thought that the drying process could be greatly sped up and the quality better preserved if all three techniques, instead of just one, were applied at once. It took him six years to work this out, but he did find a way to substantially reduce the drying time.

Always an astute marketer, he called his process anhydrous food as opposed to dehydrated, just as his food was called frosted rather than frozen. Dehydrated food was dried in eighteen hours, but anhydrous food was dried in only ninety minutes. Dried that rapidly, Birdseye claimed, the food did not have time to deteriorate and could be restored to its fresh state in between four and ten minutes.

He drew on his experience as a lamp manufacturer, using an infrared heat lamp of his own design. He gave demonstrations using his lamp and sometimes would hold up a beaker of water he had extracted from an anhydrous carrot. But the food was also heated and had air blown on it. Understanding his times, he promised that his food process would save the housewife time because the food was already partially cooked. It would also save kitchen space because a package the size of two cigarette packs would be food for a family and the small packets could be stored without risk of spoilage, saving continual trips to the market. He claimed tremendous savings for wholesalers when "five truckloads of farm produce can be processed into one truck load." The grocer, he promised, could save 80 percent of his shelf space.

Birdseye built gigantic drying machines in Gloucester. One almost twice his height was used to process carrots. He would stand on a ladder to feed the roots in. His weight sometimes less than 140 pounds, Birdseye's size made his equipment look huge.

After the outbreak of World War II, Birdseye thought his moment had come because lightweight portable non-spoiling food seemed ideal for soldiers in the field. He intensified his research in his stainless-steel-outfitted basement, installing machinery with trays of vegetables along the wall awaiting processing. By his own account, during the war he traveled thousands of miles to gather information and ideas about dehydration.

At the time, the Coast Guard was stationed on Eastern Point, patrolling for German submarines along the passage to Boston. Eleanor decided it would be a patriotic gesture to entertain these boys, have them over to the house and invite some local girls for them to meet. She hosted several such mixers. On one occasion the Coast Guard was arriving while Birdseye was absorbed in an experiment in the basement dehydrating and restoring garlic. He invited all the Coast Guard guests down to the basement and explained what he was doing and urged them to taste samples. As the girls arrived, one by one members of the Coast Guard came up from the basement, each one reeking of garlic. "Wow, were they powerful," said Lila Monell, one of the young women. "They smelled so much of garlic you couldn't get near them." It was one of those moments when Eleanor could not conceal her occasional frustration with her brilliant husband.

With wartime food shortages Birdseye was reminding people that there was lots of food out there to shoot and eat, that "food tastes are principally psychological," and that muskrat, crows, squirrels, and (was Lila right about this?) *starlings* were delicious.

When World War II ended, Birdseye officially launched his anhydrous food line in November 1945. The name Birdseye was taken seriously in the food industry. If Clarence Birdseye said that the future was dehydrated food, the press was prepared to accept that possibility. The *New York Times* ran a story with the headline "A New Page Is Turned in the History of Food." Birdseye established his Gloucester company Process Incorporated and had a number of companies signed on to distribute his anhydrous foods. Up to his old tricks, he invited two hundred food experts to lunch at the Waldorf Astoria in New York. Not until the meal was over did he reveal that the broccoli, carrots, mashed potatoes, and apple tarts were all anhydrous.

Anhydrous food did better than the "nondripping paint

brush," granted a patent on the same May day in 1947, but it was not a huge success. Soon all the new companies dealing in anhydrous food were gone. The army had used a great deal of dehydrated soups, vegetables, and stews during the war, and the soldiers had found them unsatisfactory.

Birdseye had predicted "unprecedented growth" for both frozen and dehydrated food after the war. Now he admitted that for dehydrated to keep up with frozen, the products would have to be improved. It was frozen food that was set to explode in the postwar world.

Americans bought 800 million pounds of frozen food in 1945 and 1946. *The New Yorker* ran an article on the future of frozen food that began: "Not since the appearance of the first glacier, a few aeons ago, has there been any phenomenon to compare with the frigid giant that is now looming on the horizon of the American housewife, in the shape of the frozen-foods industry." The article excitedly pointed out that there were now forty retail stores in Manhattan that sold nothing but frozen food and that twenty-two of them had opened in the past six months. Such stores were "popping up" all over the United States. Freezer manufacturing was booming, and more and more people were buying them for their homes.

By 1950, after frozen orange juice came on the market, U.S. annual frozen-food sales went to $1 billion and were projected to reach $50 billion by 1957. Frozen food now came in endless variety, though some foods still could not be successfully frozen, such as cantaloupes, grapes, lettuce, and whole onions. And others were widely recognized as inferior to fresh such as frozen eggs, cheese, and certain fish such as sole. Once the recipients of castoffs, frozen-food companies were now the customer of choice for many farmers because they paid well, bought with regularity, and gave guidance on soil and other technical issues.

Birds Eye now had stiff competition. There were more than five hundred brands of frozen food. A leading competitor was William L. Maxson, a New Yorker who made frozen meals and sold them to the Naval Air Transport Service. The navy continued to use his frozen meals after the war, serving them on navy flights. They did not sell well when offered in New York's Macy's department store, but the market was slowly growing for frozen meals. By 1950, Maxson had his first competitor, Frigid Dinners Inc. in Philadelphia. Maxson had also invented the Maxson Whirlwind, an oven to quickly thaw and cook the meals. This was the beginning of a new age. This step had been accidentally discovered in 1945 when an engineer at Raytheon working on radar discovered that resulting microwaves cooked food. In fact, Bell Laboratories had done work on this in the 1930s.

In another harbinger of the future, after the war the United States became interested in receiving frozen seafood from Asia. Small amounts of frozen seafood had been imported before the war from India and Japan. After the war the United States started importing frozen seafood from its new friend, Japan. Today frozen Asian seafood has become a major component of the American fish market.

With all the increased competition, Birds Eye, in a tradition going back to C. W. Post in the nineteenth century, fought back with aggressive and innovative advertising campaigns. In 1940 it was one of the first to advertise in color, in *Life* magazine, and it then became one of the pioneer television advertisers in the 1950s, sponsoring its own situation comedy, *Our Miss Brooks*, in which Eve Arden played a schoolteacher. Launched in 1952, it was one of the first television hits.

Birdseye always believed in the central concept of agribusiness, that through technology hunger in the world would one day vanish. He followed new industrial food ideas with great excitement. He was well aware of progress in microwaves and often mentioned it as one of the ideas of the future. Birdseye also believed in hydroponic farming. Plants grow by extracting nutrients from the soil, but if they are provided these nutrients in water, they have no need of soil, which means that fields are not necessary for growing crops. This is not a new idea. The sixteenth-century English father of science Francis Bacon wrote about it toward the end of his life, and hydroponics became very popular in the 1930s. Birdseye envisioned a New York City whose produce needs were locally supplied by rooftop hydroponic farming. He predicted that "eventually we shall learn to manufacture food from sunlight, as plants do." He also believed in the increasing use of antibiotics in food to inhibit decay.

In 1946, Birdseye telephoned twelve-year-old Don Wonson, the boy next door who was a frequent visitor, and told him to come over for pheasant. But he also told him to comb his hair and dress up. When Don went over, he found a *Look* magazine photographer in the living room. The photographer shot a picture of them in that room, with its mahogany and Persian rugs. In the caption Birdseye was quoted saying, "This is the one spot in the world we want to live in."

He was a famous and respected personality, and the world always seemed interested in hearing the next idea to come out of his kitchen or basement. He found closure to nagging unfinished business in 1941, when he finally got his Amherst degree, an honorary master of arts (Amherst still didn't give science degrees).

But his good life in Gloucester was changing. The *Sealoafer* was destroyed in a gale in 1938. He could have bought or even had built another boat, but he started to experience acute pains in his chest and was diagnosed with angina pectoris. Angina is a condition caused by an obstruction of the coronary arteries, preventing a sufficient supply of oxygen to the heart. The standard treatment at the time was nitroglycerin pills—which the body breaks down into nitric oxide, which provides the necessary oxygen—and a calmer, more sedentary life.

On the advice of his doctors, Birdseye was in search of that calmer life. There were no more whaling and no more horseback riding in West Gloucester. Bob started paying more attention to Eleanor's hobby of gardening wild plants and flowers. After they built the house, she joined the local garden club, of which she soon became president, and became increasingly passionate about filling her spacious grounds with local wild species with the help of a gifted gardener, Andrea Barletta from neighboring Rockport.

Barletta had grown up on a farm in Italy and had the knack for growing anything. It was said in Rockport that Andrea Barletta could make leaves sprout on a broomstick. He did not always earn his living gardening. For years he worked for Birdseye Electric, which is why he always called Bob "Mr. Birdseye" while Bob called him "Andy." Barletta did maintenance work, drove a truck, and frosted lightbulbs using a spray gun and a turntable to which the bulb was fastened. The machine was a Birdseye invention.

But gardening was Barletta's talent and passion. People used to drive by his home to look at and shoot photographs of his gardens with their pear trees, Japanese cherry trees, grapevines, and raspberry and blackberry bushes.

Eleanor started hosting the garden club, and one day during a meeting someone asked Bob the name of a particular fern on his grounds. Barletta remembers, "He didn't know the name. So he studied wild plants and learned the names and characteristics." Bob was probably joking when he wrote about his wife in the introduction to their 1951 coauthored gardening book, "Soon she began identifying wildflowers and ferns about which I knew nothing. That was bad for family discipline because I was supposed to be the naturalist of the clan." Whether or not it really bothered him that Eleanor was more knowledgeable than he, his ignorance on the subject certainly bothered him.

So he became deeply knowledgeable about plants, and his relaxing hobby for his new sedentary way of life started to become physically active. He began prowling Cape Ann and digging up wild species. "Hey, Andy," Bob would say, "let's go in the woods and dig up some wildflowers." All the time they were searching, Bob would question Andy about plants, his ideas, his farm in Italy.

Barletta recalled, "Mr. Birdseye and I used to go into the

southern woods near the Rockport Country Club and find plants like the cowslip and wild aster." Naturally, when out collecting plants or working in his garden, Bob wore a tie.

Reports on Bob and Eleanor's garden say there were 100 or 150 varieties of wild flowers and plants. They also planted beds of domestic flowers, especially Bob's favorite, dahlias.

The grounds were greened with Japanese pine. A friend in Nantucket, where the variety had been brought by a nineteenth-century clipper ship captain, had sent the seedlings. By 1950 the Birdseye grounds had seventy-four tall Japanese pines. Between these pines Bob and Eleanor landscaped forest soil amid chunks of the local granite and planted twenty types of ferns and other wild varieties, with winding paths for visitors to wander their woodland without stepping on the spongy, dark, loosely packed soil. Bob enthusiastically invited people to come see their garden. He started talking about how they could reproduce abundant quantities of these plants and use them to restore woodland areas in the region that had fallen into decline.

One thing he loved about gardening was that it was something he and Eleanor could do together. They didn't go whaling together, because Eleanor got terrible bouts of seasickness. But gardening was *their* project. In 1951 they published *Growing Woodland Plants*. Clarence and Eleanor G. Birdseye are listed as the authors of the book. But it appears from the writing style that while he wrote the introduction, Eleanor, the original gardener, wrote most of the very straightforward text on lady's slippers and spiderworts and other species. But when the book came out, the press rushed to interview the colorful "father of frozen food," not his shy wife, about his garden.

When he was sixty-four, Bob told an interviewer, "I am having just as much fun as I ever did. I am never bored because I am always prying into something which fascinates me."

But his mind was too active, too curious, to just stay in Gloucester forever gardening. Sooner or later something else would attract his attention. Then, too, wasn't it time for a new adventure with a new idea? The idea came to him while visiting a factory that rendered lard.

Chapter 11

Beyond the Sunset, and the Baths

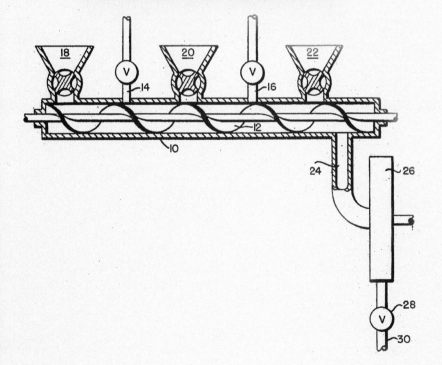

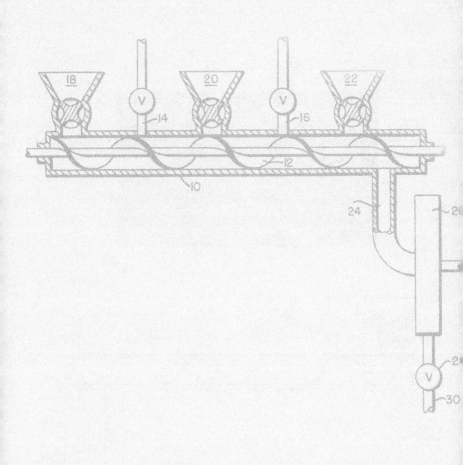

In one of the greatest poems in the English language Tennyson shows Ulysses late in life. He is getting older, has lived a domestic life for a number of years, his son is grown and doesn't need him, and he thinks of his youth and the great adventures he had. Could he not have one last adventure?

Death closes all: but something ere the end,
Some work of noble note, may yet be done,
Not unbecoming men that strove with Gods.

And so he gathers up his old comrades and leaves on one last adventure "to sail beyond the sunset, and the baths."

Most people who have known true adventure, when they start glimpsing old age and death, are consumed with the urge to stave off time with one last adventure.

Bob and Eleanor were such people. Perhaps they recalled the words of Bob's friend Wilfred Grenfell, who more than thirty years before in his autobiography had advised, "If you are reasonably resistant, and want to get tough and young again, you can do far worse than come and winter on 'the lonely Labrador.'"

Bob and Eleanor wanted to go to Labrador again—not literally, but to leave their house and garden and push off and sail beyond the sunset and the baths one last time.

Their children were grown and on their own, Kellogg and Eleanor with children. Kellogg had started a career with the Grand Union food chain, and he and his wife, Gypsy, had moved to New Jersey. Eleanor and her husband, LeMar Talbot, were in California. Henry, a geologist in search of uranium, which was one of the most highly valued resources of the 1950s, lived in Albuquerque, which set off Bob's memories of those early days on horseback in the New Mexico Territory. Henry was about to marry Ernestine LaRue Lowrey, who was from New Mexico. Only Ruth was single and still in Gloucester.

Bob and Eleanor had always been sentimental about Labrador, perhaps in the same way most couples are nostalgic about their first years of marriage. All through the years of lobster dinners and evenings of home movies and Chinese checkers, a pair of Labrador snowshoes always hung on the wall in their Gloucester home. Bob and Eleanor both frequently spoke of their time in Labrador.

The opportunity, like many opportunities, showed up unexpectedly, like an intriguing uninvited guest. Always interested in food processing, Bob was visiting a lard-rendering plant in the 1940s, and something about the process started him thinking about a better way of converting wood chips into paper pulp. Birdseye often told this story about the lard plant, but he never explained what it was about lard rendering that spoke to papermaking or, for that matter, why he was thinking about paper manufacture at all. But Birdseye was constantly thinking about industrial processes, and one thought generally led to another. Perhaps his interest in paper came from years as an innovator in food-packaging materials.

He did what he had always done: put together some financial people and some technical people and invented a new process for making paper.

By this time, when Birdseye had an idea, industrialists paid attention. The New York–based W. R. Grace and Company was interested in the Birdseye paper process. Grace was a large old international company founded by an Irish immigrant to Peru, William Russell Grace. He became involved in the guano trade, exporting bird droppings, which are rich in nitrogen and phosphorus, making it valuable material in the manufacture of gunpowder and fertilizer. Even today guano remains a prosperous South American trade. In order to ship it, a steamship company called the Grace Line was created, and it also became a commercial success. Along with guano Grace started producing and shipping sugar. The guano trade also led to the production of fertilizer. By the 1950s the company had operations in New York, London, Peru, Chile, and other places and owned the Grace National Bank, the Grace Chemical Company, and other subsidiaries around the world. As Bob explained to his daughter Eleanor and her husband, LeMar, it was "a heck of a big outfit."

Grace was interested in the possibility of applying Birdseye's process to the production of paper from cane scrap in Peru. A number of countries in the world, such as the Philippines, were suffering from acute paper shortages. Grace found other potential markets for the process in Puerto Rico, Venezuela, Argentina, and Egypt.

Grace's Peruvian sugar fields produced half a million tons of sugar annually as well as molasses and industrial-grade alcohol. This still left mountains of crushed cane stalks, called bagasse, unused. Grace made paper with it, but it was a slow, inefficient process. It took six months to dry the bagasse enough to use it for papermaking. Birdseye's process provided the possibility of using the bagasse immediately. Grace thought Birdseye could greatly improve the process and build it a new, 115-ton paper pulp factory that would be a model to sell to other countries.

And so in 1953 Bob and Eleanor left their Wyndiecote by the sea in Gloucester and moved to Paramonga, in the mountainous, cactus-studded desert of southern Peru. Grace had fifteen thousand irrigated acres of cane field, its sugar mill, and its paper plant there. It owned the town of adobe houses on unpaved roads where nine thousand people lived, most of them descendants of Incas who took the edge off their hard lives by chewing coca leaves. At one end of town was a well-kept area of forty houses where Grace executives, North Americans and Europeans, lived. Grace built a school, a movie theater, a club, and a church there.

Bob told a reporter from the *Gloucester Daily Times*, "Our own home is a single-story eight-room house, very strongly-made. It is well laid out, nicely furnished and has a large walled-in patio." They planted gardens around it with daisies, zinnias, snapdragons, bachelor's buttons, petunias, chrysanthemums, and a few crops such as parsley, lettuce, and several banana bushes. The front door was draped in fuchsia bougainvillea. The seasons barely changed, the temperature was usually in the seventies, and it could go years without raining.

There were, as Bob put it, "a couple of drawbacks." To avoid dysentery, all the water had to be boiled, the vegetables peeled, the fruit washed in disinfectant, even the milk boiled. But Bob and Eleanor weren't doing this work. The native people were very poor and worked for very little. Laborers earned eight cents an hour, while skilled workers could command fifteen cents. Maids cost $5 a month, and a really good cook could command $15 a month. The Birdseyes hired Jacinto to cook, and his wife cleaned.

This clearly wasn't Labrador. But lest there be any doubt that in the minds of Bob and Eleanor this was *their* second Labrador, one of the first things they did after settling in was begin to raise foxes. These were gray desert foxes, two of which were found by a worker on the hillside above a cane field. They were thought to be the only survivors of an unlucky brood. The two were only a few weeks old and weighed four ounces each. One slept in one of Eleanor's slippers, and another preferred a saucer. Bob and Eleanor knew how to care for foxes and were obsessed with the two, wetting their fingers with evaporated milk to feed them and giving them chopped raw meat, nursing them the way they used to when they depended on the survival of the foxes. One died, but the other, named Susie, became a center of attention. Bob and Eleanor would write pages of single-spaced typing to their children on Susie's progress.

When he got to Peru, he found the plant furnished with all the equipment he had requested. But it was still a quaint place of whistles and clanging bells and a nineteenth-century railroad that carried the cane in from the fields. Bob was working hard, up at 6:00 a.m., leaving for the plant at 6:45, working on his ideas until evening, sometimes as late as 11:00 p.m.

Grace periodically sent Bob to Puerto Rico, where the company saw tremendous opportunities for Birdseye's process. Puerto Rico at that time was an impoverished sugar-growing island. It produced mountains of bagasse and used it only as fuel to operate the mills. But Puerto Rico had American money from Operation Bootstrap, a U.S. government project to spend millions of dollars promoting industrialization on the island. Even though, ironically, part of that industrialization process was to move the Puerto Rican economy away from sugar, in the 1950s the Puerto Rican government was tremendously interested in Grace, which now had the best bagasse-to-paper process. Bob loved Puerto Rico—the flowering trees, green mountains, and blue sea. And, of course, he also reported that he liked the food.

In Paramonga, Bob had taken to making the local ceviche with red peppers he described as "twice as hot as molten steel." He thought it was a good appetizer to serve before a New England pot roast. He and Eleanor enjoyed drinking what they called "ginger con gin" until the summer of 1954, when Birdseye triumphantly announced his "most recent invention"—a coconut milk, lime juice, and gin cocktail.

He was the same Bob Birdseye, always thinking of food and writing home to say what everyone was eating. On January 10, 1954, he wrote about a trip to the coast:

> I charcoal-broiled and we four adults consumed two chickens, and in addition sundry stuffed eggs, grilled toast, potato salad, bananas, beer, coke, lemonade and for roughage, large quantities of ever drifting sand. But today we varied our picnic routine by adding surf fishing for pejerreyes—large yellow-tailed scrumptiously edible smelt-like fish.

He wrote that Eleanor was becoming proficient at catching the little fish on three hooked nylon lines, wading into the surf, a good kind of fishing for someone who gets seasick on boats.

Bob was doing a lot of fishing—and was bird shooting with the shotgun that he managed to bring in from the United States. He also started riding horses again, something he hadn't done since 1934 in West Gloucester. The terrain reminded him of the Southwest, where he'd ridden almost a half century earlier. One day fishing in about forty feet of water, he felt a powerful tug on his line and fought for thirty minutes to bring up what he thought would be an enormous fish, but it turned out to be a penguin. Once on the boat the penguin continued to fight until they tied his feet and rubber banded his sharp beak, as Birdseye said, "to keep him from filling up on Indians' toes." Billy the penguin became another household pet. Later Bob and Eleanor adopted a deer and a redheaded parrot they named Pancho, who couldn't fly because he accidentally clipped a wing while hunting. There were also geese, ducks, squabs, and guinea pigs around the house, but most of them ended up being eaten. And the Birdseyes resumed their old habit of raising Rhode Island Reds for the eggs. They had a freezer, a fifteen-cubic-foot General Electric model, to keep a supply of meat for entertaining. Appropriately, the Birdseyes brought the first freezer ever to be used in Paramonga, where most people didn't even have refrigerators. To the Birdseyes, having a freezer was important because it meant you could bring in food and not be dependent on what was available locally.

Bob was still tremendously curious about everything and wrote page after page to his children about topics from local salt making, to the enormous and curious insects he saw, to Incan pots.

Yet the Birdseyes, who didn't speak Spanish, lived a tame and very Americanized life in their comfortable foreign community. In late winter 1954 two girls from Eastern Point, Dotty Brown and Sarah Robbins, traveling through South America, stopped off to see them in Paramonga. Bob was overjoyed because he had written to Sarah to make sure she brought him Lawry's seasoning, which he loved to cook with, and she arrived with two bottles. This was a popular seasoning mixture from a Los Angeles–based chain of roast beef restaurants. It solved the problem of the local lobster. He missed his New England lobster parties. The local lobster was a freshwater crayfish with edible tails and claws that he thought tasted of mud and were "insipid." Once he had his Lawry's, he could make them appealing. He offered this recipe to his children, who, of course, didn't have the crayfish. But Bob could never resist offering a good recipe: "Shell the tails raw; split as for 'fantail shrimp'; treat with salt, pepper and Lawry's; and sauté in butter."

Dotty still remembered more than fifty years later, "We were disappointed with dinner at the Birdseyes'—roast chicken, string beans, Parker House roll, lemon meringue pie. We could have been in Gloucester!"

Bob had taught Jacinto how to make lemon meringue pie from an English cookbook. He also taught him to make coconut custard pie and sponge cake. The rolls he made himself. He also liked to make donuts. There was certainly an Americanness to their new adventure. But Dotty also remembered that they seemed very happy.

They kept thinking of Labrador and making comparisons. Shortly after arriving, Bob wrote:

Shades of Labrador! There a common winter meal
consisted of salted capelin (a smelt-like sea fish) bread
and tea. This noon I watched Indians eating the same
fodder—whole salted raw anchovies, bread and chica.
The later is the water in which dry red-kernelled corn
has been boiled, and may be either unfermented or as
heady as a good applejack.

About the time Dotty and Sarah visited, Bob and Eleanor
met a group of Christian missionaries working throughout
the region from a large organization, the Summer Institute
of Linguistics, that would later become very controversial in
Latin America. Birdseye was not uncomfortable with Christian missionaries, having a close Labrador friend, Wilfred
Grenfell, who was one. However, he noted in a letter:

Mother and I couldn't help contrasting the pioneer-
ing of this group in what only five or ten years ago
was about the wildest area in the world, with our
early experiences. Even in the summer it took us from
two to three weeks to get from any large city to our
Cartwright headquarters; in winter a 1200-mile trip
by dog team was our only means of leaving Labrador.
These folks pass from Iquitos on the Amazon to Lima
between breakfast and lunch. We had no communication with the "outside" from November 15 to June 15.
These people have constant radio communication
with Lima, Rio de Janeiro and the United States. Our
nearest doctor was 250 miles away by dogsled. They
have a resident doctor, a hospital, 3 schools, a postman

at their headquarters. We used wood-burning stoves and kerosene and melted snow for our winter water supply. They have a central power plant, refrigerators, electric stoves, and even the remotest outpost has its generator and two-way radio. Ho, hum! The world, even the jungle, does move doesn't it?

Bob believed in change and could see that there is no going back.

In the spring of 1954 Bob and Eleanor returned to the United States for almost two months. Bob had patents to register and meetings with the Grace company in New York. They had children and grandchildren to see, Gloucester to check in on, and Henry and Ernestine's wedding in Albuquerque. It was like one of those trips home from Labrador—a little business, a little shopping, and family gatherings. As evidence that they were having a second Labrador, while in New York Bob phoned contacts he hadn't used since 1910 at the American Museum of Natural History and arranged for them to ship him taxidermy equipment and materials so he could send them bird specimens. When he returned to Paramonga, he killed four types of doves and several other birds and froze them until the package arrived from the museum. In the next six months he killed and preserved sixty bird species and preserved forty mammal hides.

Late in 1955 the Birdseyes left Paramonga, abandoning Susie the fox they had so coddled, and Billy the penguin, and Pancho the parrot. They might have preferred going home to Gloucester, but Bob had work in New York with the Grace company.

They rented an ample and well-furnished apartment at the Gramercy Park Hotel opposite leafy Gramercy Park.

Dotty Brown came by to see them and thought they did not seem as happy as they had been in Peru. But when Kellogg and Gypsy visited, Gypsy was struck by how excited Bob was about his new enterprise. In his two years in Peru, Birdseye had learned how to accomplish what had been a nine-hour process in only twelve minutes. Now he was eager to market it around the world. He also retained an interest in his earlier projects, especially freezing. When the *Gloucester Daily Times* interviewed him on his paper process, he took time to caution Gloucester and its fish producers that it was essential to maintain a high level of quality in frozen fish sticks because it was the kind of product that could easily decline. "I believe the big danger to it is a drop in quality standards, which can kill a good thing almost overnight. This is not a new idea by any means, but it still holds true."

Back from Peru, he was still driven by his native enthusiasm, even though at age sixty-eight he looked thin and uncharacteristically frail. Some old friends such as his early partner in freezing, Isaac Rice, believed that Bob had ruined his health on the Grace project in Peru. He certainly hadn't lived the sedentary life his doctors had prescribed. But the doctors may have been wrong. In the 1930s, 1940s, and 1950s, when Birdseye was struggling with angina, not much was known about the heart or heart disease. Today people with Birdseye's diagnosis are told to keep active and exercise. Exercise, it turns out, releases nitric oxide, the same result as taking nitroglycerin, which opens up oxygen supply to the heart.

In any event, it seems unlikely that Birdseye would have followed doctors' orders, though he tried to. He was a bit like

Theodore Roosevelt, army colonel, naturalist, rancher, who as ex-president couldn't stay home and went on an expedition in 1914 in search of an unknown river in Brazil. He charted the river but so damaged his health that he never recovered. When the former president, aged fifty-five, was asked why he did such a foolish thing, he said simply, "It was my last chance to be a boy." Birdseye would not have passed up his South American adventure for the same reason.

From New York he assured *Time* magazine, "Still other ventures are afoot, and the days are not long enough for me to take advantage of all the opportunities I see."

On October 7, 1956, at the age of only sixty-nine, Clarence Birdseye died of heart failure in his apartment at the Gramercy Park Hotel. He had asked that rather than sending flowers to his funeral, people contribute to an Amherst College scholarship fund. He never forgot how it felt not to be able to finish your college education because you ran out of money. He asked to be cremated and have his ashes scattered in the place he loved, the sea off Gloucester.

Eleanor lived more than twenty more years. She sold the house in Gloucester and moved to Albuquerque, where Henry and Ernestine lived. She took the Labrador snowshoes with her. Ruth moved there too and got a job in the secretive research of the Sandia National Laboratories, where she was not permitted to and never did reveal the nature of her work. Eleanor studied Indian culture at the University of New Mexico and became an avid scholar, hiking to sites and ruins around the state. In 1972, Henry, an experienced pilot, was flying back to Albuquerque with five other geologists when his plane for unknown reasons crashed into a mountain in the Gila wilderness. All six died. Five years later, in 1977, at the age of eighty-eight, Eleanor died. Kellogg died in 2002, Ruth in 2003, and the daughter Eleanor in 2008.

When Bob Birdseye died in 1956, frozen food had already become a multibillion-dollar international industry. It had been developing in Britain since the end of World War II. The year Birdseye died, Giovanni Buitoni started the Italian frozen-food industry with frozen lasagna and ravioli. He had begun in the United States in 1950 and had done so well he decided to take the business back to Italy, where he sold his pasta through gelato dealers.

Eleven years after Bob Birdseye's death, he became a fictitious British character, Captain Birdseye—the respectable white-bearded old salt in a naval uniform, portrayed by actors, who promoted frozen food. Few British people even knew that there really was a Clarence Birdseye, a bald man with glasses, who had started the industry.

Today frozen food is, much the way Birdseye imagined it becoming in the 1920s, a major international business. Asian frozen-seafood exports alone account for billions of dollars in sales. Every country in the world that has a food-export business is in frozen foods. It is an essential part of modern living.

If Clarence Birdseye were to come back today, after, no doubt, telling us what everyone had been eating in the other-world, he would probably be perplexed by some of the concepts of the modern world such as the endangered species list, the whale hunting ban and marine mammal protection, limits on overfishing, the organic food movement that shuns pesticides and antibiotics, the virtues of eating locally grown foods. Many of his ideas about industrialized food are not loved today.

But in many ways he would find the world he imagined

and helped build. He would be gratified to see his reflecting lightbulbs in common use. Though he is not remembered for it, it was one of his most successful inventions. Poor sugar-producing countries do get paper from sugar bagasse. It is, as he predicted, a world in which food transcends geography and climate: any food is available anywhere at any time of year. Frozen food is commonplace, a large part of stores and super-markets and other retail food outlets, and many people keep their food in a home freezer, place it in their home microwave, and prepare their dinners in a few minutes.

But perhaps the more important thing about Clarence Birdseye was his ability to live life as an adventure. Curiosity is the one essential ingredient to an adventurous life. Isn't an original life one of the greatest inventions?

Acknowledgments

A big thanks to all of my friends in Gloucester for their support and interest. A special thanks to Sarah Dunlap in the Gloucester city archives for working with me and tirelessly sifting through records to separate fact from fiction. Thanks to those two great Gloucester institutions, the Cape Ann Museum and the Sawyer Free Library. Also much thanks to Maggie Rosa, who generously opened her files and address book and pointed me in several directions, and to other Gloucester people, including Dotty Brown, Lila Monell, Don Wonson, Janis and John Bell, Melissa Palladino, and so many others from the world's best town.

I could never have done this book without the kindness, generosity, and openness of the Birdseye family—Michael, Kelly, Gypsy, and especially Henry, who finally looked in his attic.

Thanks to my brother Paul for helping me understand the treatment of a heart condition in Birdseye's day. Thanks to Dylan Hixon for talking me through the freezing and melting properties of salt. Thanks to Jessie Cohen for first talking about this project, to the Alfred P. Sloan Foundation for its support, and especially to my great editor, Gerry Howard, for all his help and advice, enthusiasm and guidance. And thanks to Susan Birnbaum Fisher for all her help. And of course, thanks to the great New York Public Library. Where else could you look up Clarence Birdseye's 1886 address in five minutes?

Thanks to my agent, Charlotte Sheedy, for being my advocate and friend. And thanks to my tireless research assistant, Talia Kurlansky, and to Marian Mass, who are always in my heart.

Bibliography

Books

Amherst College. *The Olio.* Amherst, Mass.: Amherst College, 1908.

Birdseye, Clarence F. *American Democracy Versus Prussian Marxism: A Study in the Nature and Results of Purposive or Beneficial Government.* New York: Fleming H. Revell, 1920.

Birdseye, Clarence, and Eleanor Birdseye. *Growing Woodland Plants.* New York: Oxford University Press, 1951.

Carlton, Harry. *The Frozen Food Industry.* Knoxville: University of Tennessee Press, 1941.

Davies, David, and Alf Carr. *"When It's Time to Make a Choice": 50 Years of Frozen Food in Britain.* Grantham, Lincolnshire: British Frozen Food Federation, 1998.

De Sounin, Leonie. *Magic in Herbs.* With an introduction by Miriam Birdseye. New York: Gramercy, 1941.

Dolin, Eric Jay. *Fur, Fortune, and Empire: The Epic History of the Fur Trade in America.* New York: W. W. Norton, 2010.

Evans, Harold. *They Made America: From the Steam Engine to the Search Engine: Two Centuries of Innovators.* With Gail Buckland and David Lefer. New York: Little, Brown, 2004.

Foley, John. *The Food Makers: A History of General Foods Ltd.* Banbury, Oxfordshire: General Foods, 1972.

Freidberg, Susanne. *Fresh: A Perishable History.* Cambridge, Mass.: Belknap Press of Harvard University Press, 2009.

Fucini, Joseph, and Suzy Fucini. *Entrepreneurs: The Men and Women Behind Famous Brand Names and How They Made It.* Boston: G. K. Hall, 1985.

Garland, Joseph E. *Eastern Point: A Nautical, Rustical, and More or Less Sociable Chronicle of Gloucester's Outer Shield and Inner Sanctum, 1606–1990.* Beverly, Mass.: Commonwealth, 1999.

Grenfell, Wilfred Thomason. *Adrift on an Ice-Pan.* Boston: Houghton Mifflin, 1909.

———. *A Labrador Doctor: The Autobiography of Wilfred Thomason Grenfell.* Boston: Houghton Mifflin, 1919.

———. *Tales of the Labrador.* Boston: Houghton Mifflin, 1916.

Hammond, John Hays. *The Autobiography of John Hays Hammond.* 2 vols. New York: Farrar and Rinehart, 1935.

Harden, Victoria A. *Rocky Mountain Spotted Fever: History of a Twentieth-Century Disease.* Baltimore: Johns Hopkins University Press, 1990.

Kerr, J. Lennox. *Wilfred Grenfell: His Life and Work.* New York: Dodd, Mead, 1959.

Leonard, John William, ed. *Who's Who in New York City and State.* New York: L. R. Hamersly, 1907.

———. *Woman's Who's Who of America: A Biographical Dictionary of Contemporary Women of the United States and Canada, 1914–1915.* New York: American Commonwealth, 1914.

Parkes, A. S., and Audrey U. Smith, eds. *Recent Research in Freezing and Drying.* Springfield, Ill.: Charles C. Thomas, 1960.

Price, Esther Gaskins. *Fighting Spotted Fever in the Rockies.* Helena, Mont.: Naegele Printing, 1948.

Rothe, Anna, ed. *Current Biography: Who's News and Why, 1946.* New York: H. W. Wilson, 1947.

Shachtman, Tom. *Absolute Zero and the Conquest of Cold.* Boston: Houghton Mifflin, 1999.

Tressler, Donald. *The Memoirs of Donald K. Tressler.* Westport, Conn.: Avi, 1976.

Tressler, Donald, and Norman W. Desrosier, eds. *Fundamentals of Food Freezing.* Westport, Conn.: Avi, 1977.

Tressler, Donald, and Clifford F. Evers. *The Freezing and Preservation of Foods.* 2 vols. Westport, Conn.: Avi, 1957.

U.S. Fisheries Association. Proceedings of the Sixth Annual Convention of the United States Fisheries Association. Atlantic City, New Jersey. United States Fisheries Association, 1924.

U.S. Patent Office. *Selected Patents of Clarence Birdseye.*

Williams, E. W. *Frozen Foods: Biography of an Industry.* Boston: Cahner's, 1968.

Newspaper, Magazine, and Journal Articles

Bald, Wambly. "Lion a Tidbit to Frozen-Food Genius." *New York Post*, December 20, 1945.

Birdseye, Clarence. "The Birth of an Industry." *Beaver*, September 1941 (magazine of the Hudson's Bay Company).

———. "Camping in a Labrador Snow-Hole." *Outing*, November 1913.

———. "The Gravity Froster." *Refrigerating Engineering*, November 1940.

———. "Hard Luck on the Labrador." *Outing*, April–September 1915.

———. "If I Were Twenty-one." *American Magazine*, February 1951.

———. "Looking Backward at Frozen Foods." *Refrigerating Engineering*, November 1953.

———. "Postwar Problems of the Frozen Food Industry." *Meals for Millions.* Final Report of the New York State Joint Legislative Committee on Nutrition, 1947.

———. Some Common Mammals of Western Montana in Relation to Agriculture and Spotted Fever. U.S. Department of Agriculture, Farmers' Bulletin, March 9, 1912.

———. "We Can Always Eat Crow." *American Magazine*, July 1943.

Burbank, Russell. "Frozen Food Site to Close." *Boston Globe*, July 18, 1965.

Burton, L. V. "Birdseye Demonstrates New Twenty-Plate Froster." *Food Industries*, November 1941.

Carlson, Scott. "Birds Eye Foods—A Short History of the Frozen Food Section." *Washington Business Magazine*, May/June 2004.

Erkkilla, Barbara. "Birdseye Produces Paper Pulp from Sugar Cane Stalk." *Gloucester Daily Times*, September 6, 1955.

———. "Quality the Key to 'Stick' Future Birdseye Believes." *Gloucester Daily Times*, September 6, 1955.

Farrell, Morgan. "Quick Food Freezing Process Devised to Aid the Housewife." *New York Times*, February 14, 1932.

Harris, Herbert. "The Amazing Frozen-Foods Industry." *Science Digest*, February 1951.

Henshaw, Henry Wetherbee, and Clarence Birdseye. "The Mammals of Bitterroot Valley, Montana, in Their Relation to Spotted Fever." U.S. Department of Agriculture, Bureau of Biological Survey, Circular No. 82, August 3, 1911.

Kahn, F. J., Jr. "The Coming of the Big Freeze." *The New Yorker*, September 14, 1946.

Kenyon, Paul. "About the Birdseyes . . . and More." *Gloucester Daily Times*, April 14, 1978.

———. "Eleanor Birdseye, a Working Wife." *Gloucester Daily Times*, March 5, 1977.

Lee, Frank A., Robert F. Brooks, A. M. Pearson, John I. Miller, and Frances Volz. "Effect of Freezing Rate on Meat." *Journal of Food Science*, September 15, 1919.

Mallon, Winifred. "More Food Patents Won by Birdseye." *New York Times*, May 3, 1947.

Moore, Josiah J. "Time Relationships of the Wood Tick in the Transmission of Rocky Mountain Spotted Fever." *Journal of Infectious Diseases*, April 12, 1911.

Morgan, T. H. "Breeding Experiments with Rats." *American Naturalist* 43 (1909): 183–85.

Nickerson, Jane. "New and Better Way to Process Food by Anhydration Announced by Birdseye." *New York Times*, November 14, 1945.

Pennington, Mary E. "Refrigerated Trucks Essential in Sale of Frozen Foods." *Refrigerating Engineering*, November 1940.

Stiebeling, Hazel K., and Miriam Birdseye. *Adequate Diets for Families with Limited Incomes*. U.S. Department of Agriculture, Miscellaneous Publication 113, April 1931.

Tressler, Donald, Clarence Birdseye, and William T. Murray. "Tenderness of Meat." *Industrial and Engineering Chemistry* 24, no. 2 (1932).

Chronology of Unsigned Articles

Scientific American, April 21, 1860, 267 (Henry Underwood).

"Underwood Cotton-Leather Belting." *Electrical Engineer*, September 2, 1891, 271.

"Lieut. Birdseye Wounded: Son of New York Lawyer Fighting with Canadians in France." *New York Times*, October 30, 1916.

"The Looting of the Pittsburgh Life and Trust: A Narrative of Astounding Effrontery and Rascality." *Insurance Press*, May 9, 1917.

"Birdseyes Get Jail Terms." *New York Times*, March 6, 1920.

"Postum to Get Freezing Process." *New York Times*, May 8, 1929.

"Postum Buys General Seafoods and Its Rights." *Gloucester Daily Times*, May 9, 1929.

"New Postum Name on Stock Board Today." *New York Times*, July 25, 1929.

"For the First Time Anywhere! The Most Revolutionary Idea in the History of Food Will Be Revealed in Springfield Today." *Springfield Union*, March 6, 1930.

"Mayor Praises Clarence Birdseye for Work in Refrigeration Field." *Boston Globe*, July 8, 1931.

"Sachs Tells Story of $12,000,000 Loss." *New York Times*, May 21, 1932.

"Acquires Birdseye Electric." *New York Times*, December 13, 1939.

"Continuous Quick Freezer Developed by Birdseye." *Food Industries*, September 1940.

"Dinner, Frozen or Dried." *Newsweek*, November 19, 1945.

"Sylvania to Take Over Wabash." *New York Times*, December 9, 1945.

"Meet Mr. Birdseye—He Had a $22,000,000 Idea." *Look*, April 30, 1946.

"Birdseye Changes Pace, Now Raises Wildflowers." *Cape Ann Summer Sun*, August 17, 1950.

"Father of Frozen Foods Dies: Clarence Birdseye Dead at 69." *Gloucester Daily Times*, October 8, 1956, 1.

"Clarence Birdseye Is Dead at 69; Inventor of Frozen Food Process." *New York Times*, October 9, 1956.

"Clarence Birdseye Funeral Held Here." *Gloucester Daily Times*, October 10, 1956.

"The Inquisitive Yankee." *Time*, October 22, 1956.

"Tribute to Clarence Birdseye, the Father of Frozen Foods." *Quick Frozen Foods*, March 1960.

"Hooked on Fish." *Forbes*, September 1, 1969.

"Miss Marjorie Merriweather Post Is Dead at 86." *New York Times*, September 13, 1973.

"Growing Up a Birdseye—A Daughter Remembers Her Dad." *Link* 9, no. 3 (2005).

Cold Storage and Ice Trade Journal. Ice Trade Journal 38 (July–December 1909). Reprints from the collection of the University of Michigan Library, 2010.

WEB SITES

The Internet is full of inaccurate information on Birdseye, most of it feeding off each other. But there are a few useful sites:

www.amherst.edu (for information about the Birdseyes at Amherst)
www.birdseyefoods.com (for Birds Eye corporate history)
www.foodreference.com (for a food time line)
www.library.hbs.edu (for a list of deals involving General Foods)

Illustration Credits

Page 4

Top left: Courtesy of the Birdseye family.

Top right: World-Telegram Photo/C.M. Stieglitz. The *New York World-Telegram* and the *Sun Newspaper* Photograph Collection (Library of Congress).

Bottom: From the *Springfield Sunday Union and Republican,* March 2, 1930; p. 20 A. Birds Eye® is a registered trademark of Pinnacle Foods Group LLC.

Page 5

Top: Cape Ann Museum, Gloucester, MA, USA.

Center: Courtesy of the Birdseye family.

Bottom: Courtesy of the Birdseye family.

Page 6

Top left: © NMPFT/Hulton-Getty/Science & Society Picture Library. All rights reserved. Used with permission.

Top right: Cape Ann Muesum, Gloucester, MA, USA. Birds Eye® is a registered trademark of Pinnacle Foods Group LLC.

Bottom: Courtesy of the Birdseye family. Birds Eye® is a registered trademark of Pinnacle Foods Group LLC.

Page 7

Top: Courtesy of the Birdseye family.

Bottom left: Courtesy of the Birdseye family.

Bottom right: Courtesy of the Birdseye family.

Page 8

Top: Courtesy of the Birdseye family.

Bottom: Cape Ann Museum, Gloucester, MA, USA.

Index

About the Author

Mark Kurlansky is the *New York Times* bestselling author of many books, including *The Food of a Younger Land, Cod: A Biography of the Fish That Changed the World, Salt: A World History, 1968: The Year That Rocked the World,* and *The Big Oyster: History on the Half Shell.* He lives in New York City.

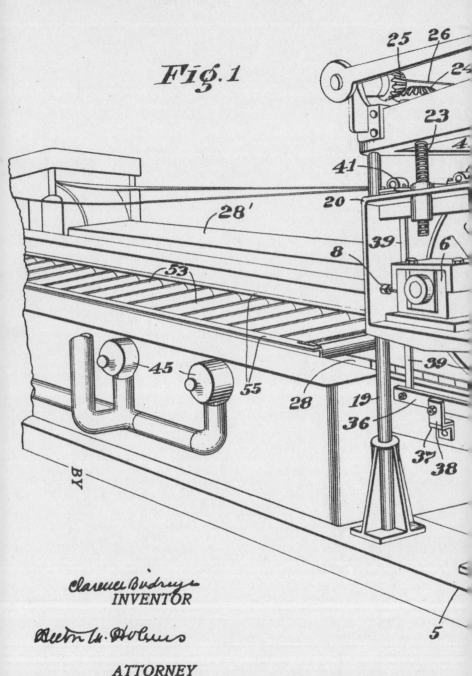

Aug. 12, 1930.

Fig.1

25   26   24

23

4

41

20

39   6

8

53

39

45

55

28'

28   19

36

37   38

BY

Clarence Birdseye
INVENTOR

Beeton W. Holmes

ATTORNEY

5